ECONOMIC DEVELOPMENT IN RURAL AREAS

The Dynamics of Economic Space

Series Editor: Christine Tamásy, University of Vechta, Germany

The IGU Commission on 'The Dynamics of Economic Space' aims to play a leading international role in the development, promulgation and dissemination of new ideas in economic geography. It has as its goal the development of a strong analytical perspective on the processes, problems and policies associated with the dynamics of local and regional economies as they are incorporated into the globalizing world economy. In recognition of the increasing complexity of the world economy, the Commission's interests include: industrial production; business, professional and financial services, and the broader service economy including e-business; corporations, corporate power, enterprise and entrepreneurship; the changing world of work and intensifying economic interconnectedness.

Other titles in this series

Regional Resilience, Economy and Society
Edited by Christine Tamásy and Javier Revilla Diez
ISBN 978-1-4094-6848-6

Local Food Systems in Old Industrial Regions
Edited by Neil Reid, Jay D. Gatrell, and Paula S. Ross
ISBN 978-1-4094-3221-0

Industrial Transition
Edited by Martina Fromhold-Eisebith and Martina Fuchs
ISBN 978-1-4094-3121-3

Missing Links in Labour Geography
Edited by Ann Cecilie Bergene, Sylvi B. Endresen and Hege Merete Knutsen
ISBN 978-0-7546-7798-7

Globalising Worlds and New Economic Configurations
Edited by Christine Tamásy and Mike Taylor
ISBN 978-0-7546-7377-4

Agri-Food Commodity Chains and Globalising Networks
Edited by Christina Stringer and Richard Le Heron
ISBN 978-0-7546-7336-1

Economic Development in Rural Areas

Functional and Multifunctional Approaches

Edited by

PETER DANNENBERG
University of Cologne, Germany

ELMAR KULKE
Humboldt-Universität zu Berlin, Germany

Routledge
Taylor & Francis Group

LONDON AND NEW YORK

First published 2015 by Ashgate Publishing

Published 2016 by Routledge
2 Park Square, Milton Park, Abingdon, Oxon OX14 4RN
711 Third Avenue, New York, NY 10017, USA

First issued in paperback 2018

Routledge is an imprint of the Taylor & Francis Group, an informa business

British Library Cataloguing in Publication Data
A catalogue record for this book is available from the British Library.

The Library of Congress has cataloged the printed edition as follows:
Economic development in rural areas : functional and multifunctional approaches / by Peter Dannenberg and Elmar Kulke.
    pages cm. -- (The dynamics of economic space)
 Includes bibliographical references and index.
 ISBN 978-1-4724-4481-3 (hardback) -- ISBN 978-1-4724-4482-0 (ebook) -- ISBN 978-1-4724-4483-7 (epub) 1. Rural development. 2. Economic development. 3. Rural development--Case studies. 4. Economic development--Cases studies. 5. Agricultural productivity--Case studies. I. Dannenberg, Peter (Professor) II. Kulke, Elmar.
 HN49.C6E26 2015
 307.1'412--dc23

2014045112

ISBN 13: 978-1-138-54679-0 (pbk)
ISBN 13: 978-1-4724-4481-3 (hbk)

Contents

PART III   ALTERNATIVE FUNCTIONS FOR RURAL AREAS

PART IV   CONCLUSION

List of Figures

List of Tables

Notes on Contributors

Dr Adam Czarnecki is Postdoctoral Researcher at the Karelin Institute, University of Eastern Finland, Finland.

Dr Anika Trebbin is Senior Lecturer at the Institute of Geography at the Philipps-Universität Marburg, Germany.

Dr Anna Kołodziejczak is Professor at the Institute of Socio-Economic Geography and Spatial Management at the Adam Mickiewicz University, Poznan, Poland.

Dr Barbara Maćkiewicz is Senior Lecturer at the Institute of Socio-Economic Geography and Spatial Management at the Adam Mickiewicz University, Poznan, Poland.

Becca B.R. Jablonsky is Researcher for the Department of City and Regional Planning as well as Predoctoral Fellow at the US Department of Agriculture's National Institute of Food and Agriculture, Cornell University, Ithaca, USA.

Diane M. Miller is Assistant Vice-President of Federal Relations at the University of Toledo, USA.

Dr Elmar Kulke is Professor of Economic Geography at the Department of Geography at the Humboldt-Universität zu Berlin, Germany.

Dr Ewa Kacprzak is Senior Lecturer at the Institute of Socio-Economic Geography and Spatial Management at the Adam Mickiewicz University, Poznan, Poland.

Dr Frank Calzonetti is Professor at the Institute of Geography and Vice President for Government Relations at the University of Toledo, USA.

Dr Gilbert Nduru is Professor of Geography and Human Ecology at Karatina University, Kenya.

Dr Heide Hoffmann is Senior Lecturer of Agroecology and Organic Farming at the Humboldt-Universität zu Berlin, Germany.

Dr Javier Revilla Diez is Professor at the Institute of Economic and Cultural Geography, Leibniz University of Hannover, Germany.

Dr Jay D. Gatrell is Dean of the College of Graduate and Professional Studies and a member of the Geography faculty at Bellarmine University, USA.

Dr Kai Mausch is Scientist of Economics at the International Crops Research Institute for the Semi-Arid Tropics, Nairobi.

Dr Krystian Heffner is Professor at the Institute of Geography at the University of Economics in Katowice, Poland.

Dr Luisa Vogt is Senior Lecturer at the Institute of Geography at the South Westphalia University of Applied Sciences, Soest, Germany.

Dr Marcus Mergenthaler is Professor of Agricultural Engineering at the South Westphalia University of Applied Sciences, Soest, Germany.

Dr Markus Hassler is Professor at the Institute of Geography at the Philipps-Universität Marburg, Germany.

Dr Martin Franz is Senior Lecturer at the Institute of Geography at the Philipps-Universität Marburg, Germany.

Dr Neil Reid is Professor at the Institute of Geography and Planning at the University of Toledo, USA.

Nithya Vishwanath Gowdru is Researcher at the Institute of Agricultural Economics and Social Sciences at the Humboldt-Universität zu Berlin, Germany.

Dr Peter Dannenberg is Professor of Anthropogeography at the Department of Geography, University of Cologne, Germany.

Ravi Nandi is Researcher at the Institute of Agricultural Economics and Social Sciences at the Humboldt-Universität zu Berlin, Germany.

Dr Sabine Panzer-Krause is Senior Lecturer of Human Geography and Economic Geography at the University Hildesheim, Germany.

Dr Wolfgang Bokelmann is Professor of Agricultural Economics at the Humboldt-Universität zu Berlin, Germany.

PART I
Introduction

Chapter 1

Introduction: Dynamics in Rural Development Beyond Conventional Food Production

Peter Dannenberg and Elmar Kulke

Introduction

This book outlines current economic dynamics in rural areas with a focus on recent developments in agricultural production which go beyond the classical conventional food production as it emerged e.g. in Europe and North America after World War II. Starting from a general discussion on the changing importance of agricultural production and the development of structural weaknesses we broaden our perspective by taken into account the changing role and functions of agricultural production systems and rural areas in the context of global and local changes. Here, we discuss the conceptual works on structural weaknesses of rural areas, regional production systems and value chains as well as functional and multifunctional approaches and present case studies of recent developments from Belgium, Germany, India, Kenya, Poland, the United States, and Vietnam.

Structural Weaknesses of Rural Areas

The variety of definitions of rural areas and their defining characteristics is quite broad in nature. Typical constituting characteristics can be seen in contrast to urban or suburban areas, such as lower population and building density. The role of agricultural production as a constitutive element of rural areas is outlined less prominently (see e.g. Isserman, 2005; McCarthy, 2005). In industrialized societies, agricultural production as a basis for growth and employment has declined over the past decades. This is discussed in various works on structural changes and weaknesses in rural areas (see e.g. Goodman and Watts, 1994; Anríquez and Stamoulis, 2007; Sedlacek et al., 2009; Dannenberg, 2010). From the early 1950s to the mid-1980s in particular, rural areas in the global North were often marked by concentration processes in agriculture, increasing mechanization, more intensive biochemical input use and increasing specialization (productivism;

Ilbery and Bowler, 1998). While this did lead to an overall increase in production, it also contributed to a loss of employment in agriculture which was formerly the backbone of the socio-economic structure of these areas. Today, the share of core agricultural production in terms of gross domestic product and employment in rural areas in most industrialized nations has lost its dominant position and often lies well below 10 per cent (Dannenberg and Kulke, 2005; Anríquez and Stamoulis, 2007).

This – together with other developments – has led to high unemployment rates and a decreasing population, especially in peripheral rural areas, whereas positive migrant flows and economic activities cumulate in larger cities and commuter belts. While typically young and well educated people leave rural areas, the elderly and retired as well as the less qualified workers stay behind. The combination of an already low population in rural areas together with outmigration makes investing in infrastructure in these regions very expensive in relation to population density. Many schools and kindergartens close down for lack of demand, which hinders young families from migrating to such regions. The result is typically a negative cumulative process of population decline, loss of jobs, loss of infrastructure (including public areas of transport, health, but also private infrastructure like food retailing) and further outmigration (see Figure 1.1; Wießner, 1999; Sedlacek et al., 2009).

On the other hand also in industrialized and highly developed countries – even without having a larger direct socio-economic impact anymore – agricultural production is still shaping the structure of the landscape of rural areas and can be seen as a crucial factor for the development of various activities and functions (e.g. housing, recreation, and environment).

In rural areas of the global South – where rapid industrialization has not taken place so far – agricultural production is, even today, frequently the most important factor for socio-economic development (Anríquez and Stamoulis, 2007). However, agricultural production in the global South is often marked by limited market opportunities. Furthermore, producers all over the world are increasingly facing international competition (Wilson, 2001; D'Hease and Kirsten, 2006; Woods, 2012). As a result, rural areas in the global South are often marked by severe structural deficits. This is especially the case in sub-Saharan Africa, where large parts of the rural population are facing poverty and vulnerability with no stable source of income, no reliable access to food, clean drinking water, health services and education (D'Hease and Kirsten, 2006). Opportunities to secure a decent livelihood through employment and self-employment apart from agriculture are in many cases non-existent, insufficient or unstable.

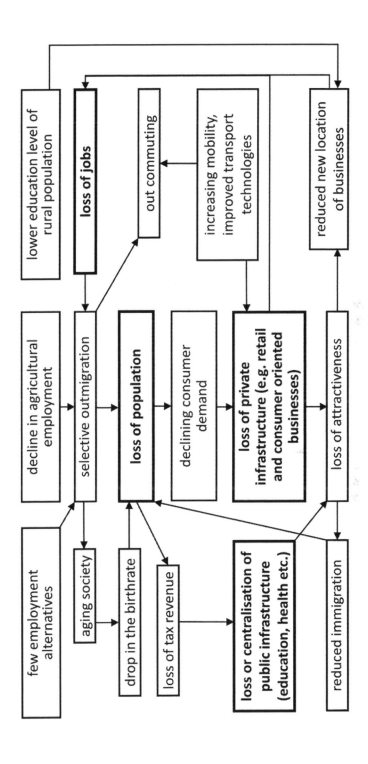

Figure 1.1   Typical cumulative process of structural weaknesses in rural areas
Source: Own design; Wießner, 1999: 301; Sedlacek et al., 2009.

Current Dynamics in Regional Agrarian Systems and Value Chains

Given the challenges and structural weaknesses outlined above, there is much discussion in the international arena as to how rural development can be sustainably achieved. In the agricultural sector, this includes research on how regional agrarian systems can either gain or maintain a competitive position in both national and international markets. In Economic Geography and related disciplines this is reflected in a number of studies conducted to analyse the impact of regional production systems or clusters in terms of their function; as a source of business knowledge exchange and as a means to create and maintain competitive advantages (Porter and Bond, 1999; Bathelt, 2005; Dannenberg and Kulke, 2005). Such regional agrarian systems do not only consist of agricultural producers but also suppliers (e.g. of seeds, chemicals, and equipment), buyers (e.g. processors and wholesalers) and public and private services (e.g. extension and finance) which are interlinked with each other. A mutual interaction and cooperation between these actors can lead to an exchange of experiences, enlarges product competences and therefore the competitiveness of the involved units. Given the large number and variety of actors involved, a broad regional agrarian system incorporates a significant volume of economic activities and therefore has the potential to substantially increase the economic performance of the whole rural area, assuming it can make use of its advantages (Dannenberg and Kulke, 2005; Morrison and Rabellotti, 2009).

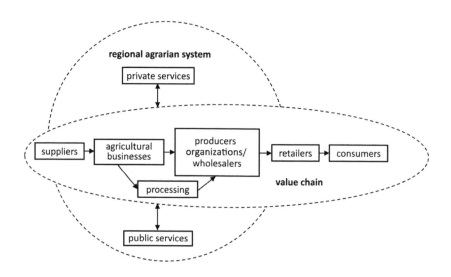

Figure 1.2   Typical connections between the regional agrarian system and the value chain

Source: Dannenberg and Kulke, 2005.

Besides the local linkages, agricultural units are more and more integrated and influenced by supraregional commodity chains (see Figure 1.2). The interrelations within the chain – based on the flow of products, the exchange of information and power relations – strongly influence the form and intensity of agricultural production. During the last decade, supraregional interrelations have markedly increased and even remote areas of developing economies have been influenced by these linkages. The recent scientific discussion on value chains considers these interrelations in the concepts of global commodity chains (GCC), of global value chains (GVC) and of production networks (GPN):

- The global commodity chain approach (Gereffi, 1996, 2001) focuses on material/immaterial linkages (e.g. the flow of goods and information) and is especially useful for analysing power relations in the chain; the agricultural sector is often considered to be strongly influenced by powerful buyers such as wholesalers, food processing and supermarket chains. Some of the chapters in this volume analyse these power relations, as agriculture seems to be increasingly dependent on external standards (e.g. social and environmental standards), on the purchasing power of retail chains and on changes in consumer behaviour, rather than on product standards alone.
- The global value chain approach (Gereffi et al., 2005; Gibbon et al., 2008) considers the different forms of coordination within the chain. The complexity of knowledge transfer, the codification of knowledge and the capabilities of the actors involved are important elements which influence the different forms of coordination. For agricultural production, the flow of information is increasingly important; it presents possibilities to upgrade processes (in other words to innovate to increase the value of products, processes or of functions). Upgrading can increase farmer incomes and may open new market opportunities for specialized products; some of the chapters discuss the potential and success factors of these.
- The global production network approach (Henderson et al., 2002; Coe et al., 2008) discusses the vertical linkages within the chain in addition to the horizontal relations of actors in the regional environment (such as universities, financial, state, private, educational, research, and development institutions). Three elements of analysis characterize this comprehensive approach: embeddedness, power, and value. Agricultural production may be strengthened by establishing collective power or institutional power based on horizontal linkages; developing networks or realizing diversification (e.g. in organic production or agritourism).

Dynamics in the Functions of Rural Areas and Multifunctional Approaches

Given the loss of importance of the core agricultural production (farming) – as a direct factor for economic development and employment, especially in the global North but also in the global South – rural areas are increasingly regarded as providers of other functions for society. From an analytical perspective, common classifications of the functions of rural areas include agriculture, leisure and recreation, ecosystem services, locations for businesses, infrastructure, and housing (see e.g. Goodman and Watts, 1994; Henkel, 2004; Pollermann et al., 2013).

This analytical approach still seems viable for a general categorization of the various functions of rural areas. However, as discussed above, today rural areas and their functions are affected by different global and local developments and frameworks including (D'Hease and Kirsten, 2006; Palang et al., 2006; Dannenberg, 2009; Hughes, 2009; Pacione, 2009):

- a growing urbanization of the world population;
- globalization processes;
- technological progress;
- dependencies and power asymmetries between urban centres and actors on the one hand and rural areas and actors on the other hand;
- changing political paradigms and regulatory regimes (e.g. the reforms of the agricultural regulations of the European Union, GATT/World Trade Organization (WTO) and the NAFTA or the still ongoing structural changes in Eastern Europe after the political transformation);
- changing consumer behaviour (e.g. required transparency in food consumption);
- changing demands of society (e.g. a growing environmental awareness); and
- demographic changes.

As a result, over the last decades new functional and multifunctional approaches to food production and agriculture but also to rural landscapes have been topics of great discussion and interest in the scientific community as well as among politicians, planners and businesses (FAO, 2000; McCarthy, 2005; Rigg, 2006; Wüstemann et al., 2008). These works take the outlined overlapping and combined functions of rural areas into account. In agriculture this includes e.g. the discussion of how far the shift from productivist production to the wide-ranging diversity that now exists within the productivist/post-productivist spectrum (seen for example in the rise of organic farming and alternative food networks; Wilson, 2001) took place.

In this context, the contribution of agriculture beyond commodity production to other functions and economic sectors is discussed. As stated by the Food and Agriculture Organization of the United Nations (FAO, 2000), the most important role of agriculture remains the production of food. However, agriculture and related land use activity can also deliver a wide range of non-food goods and services (including ecosystem services), influence the natural resource system, shape social and cultural systems and contribute significantly to economic growth. Such contributions are for example identified in the practice of sustaining attractive rural landscapes as an amenity and cultural heritage tribute (which are important for the tourism and housing sector), protecting biodiversity, generating employment and generally contributing to the viability of rural areas. In this way the concept of multifunctionality can be used as an integrated framework to interrogate contemporary rural dynamics which go beyond a simple classification of actors and their functions (McCarthy, 2005; Wüstemann et al., 2008).

Also in Western politics and planning a shift of agricultural politics from a commodity output orientated productivist focus to a broader perspective on different values, functions and possible activities of rural areas has taken place. This can be seen for example in the progressive withdrawal of production orientated state subsidies, growing environmental regulations, the promotion of endogenous development and building the capacity of rural people in different economic sectors, as practiced for example in the LEADER initiative of the European Union (Wilson, 2001; Evans et al., 2002; Shucksmith, 2010; Domon, 2011). In developing countries the FAO, the International Fund for Agricultural Development (IFAD), and other organizations underline the role of food security which goes beyond the pure production of food (which could be limited for example on export-orientated cash crops) but also provides stable availability and access to food for the local rural population (FAO, 2000). However, also in the global South, further multifunctional and post-productivist perspectives are gaining importance. For example organic farming, environmental protection and agricultural tourism are expanding (Raynolds, 2004; Rigg, 2006).

However, even though new functional and multifunctional approaches have been critically and controversially discussed for years it still remains unclear how far such new rural functions or the shift to new types of production will result in a better or more sustainable development of rural areas. A more in-depth, scientific analysis of the developments and perspectives of these functions is needed.

Our Contributions in This Volume

The aim of this book is therefore to analyse, explain, and assess ongoing changes and dynamics in rural development from a functional perspective. The preceding book in this series, *Regional Resilience, Economy and Society: Globalizing Rural Spaces* (Tamásy and Diez, 2014), took an actor-based perspective, focusing on the behaviour of rural actors in a globalizing world. This book develops an additional understanding in how far traditional functions of rural areas are changing, new functions in rural areas are developing and multifunctional approaches, actors and developments are gaining importance. Furthermore it outlines new approaches in policy, practice and theory. Here we start from a narrow perspective on current approaches and challenges of agriculture and afterwards broaden our view to further related developments in rural areas.

While rural geography was in the past mainly focused on rural areas of industrialized anglophone countries (see also the critique by McCarthy, 2005), this book aims to discuss developments in rural areas in different parts of the world. Following this introduction (Part I) the chapters can be divided into two main parts: Part II and Part III.

Part II brings together different case studies which provide examples of ongoing processes and new developments that are taking place within the agricultural food production function. Here we outline ongoing external and internal factors which are shaping rural food production systems and the related challenges. In Chapter 2, Dannenberg and Nduru discuss the influence of horizontal linkages and embeddedness in regional agrarian systems of export oriented horticultural farmers in Kenya, along with the impact of these in terms of competitiveness in international value chains. Chapter 3, by Vishwanath Gowdru et al., also focuses on small scale farmers in the global South, examining organic and conventional tomato farmers and the factors which influence their market access. Both examples show that the simple production of food, even in developing countries, is often not enough to secure entrance into the markets of choice. Farmers are instead required to fulfil special standards set by large retail chains, in addition to complying with consumer demand. This is often manifested in product and process standards including environmental (e.g. chemical and water usage), social (e.g. worker protection), and economic aspects (e.g. delivery terms and conditions). These requirements are particularly difficult for small scale farmers with low financial capacity to fulfil.

Regarding the global North, Miller et al. (Chapter 4) discuss the challenges faced by traditional farming systems in the United States and Belgium to overcoming outdated business practices and learning to utilize new technology. Jablonski (Chapter 5) explores the literature on the role of rural–urban economic linkages on rural resilience, and the evidence for relocalized food systems as a viable strategy to support rural economic development. Using a case study from Germany, Vogt and Mergenthaler (Chapter 6) analyse short food supply chains and identify typical success factors and bottlenecks as well as starting points for policy advice.

In Part III we give examples of rural functions which go beyond food production, even though they are often linked to it. Here we offer a differentiated discussion of practiced alternatives for rural development including leisure and recreational functions, environmental functions, housing functions and different business location functions. This is introduced by Panzer-Krause (Chapter 7), who analyses the potential of rural small and medium enterprises in the renewable energy market in Germany. Another German case study, analysing wholesale cooperation as an alternative way of securing local supply in rural areas of the West German federal state Hesse, is presented by Trebbin, Franz, and Hassler (Chapter 8).

Looking towards the other side of the River Oder to Poland, Kacprzak and Maćkiewicz (Chapter 9) examine organic agritourism; an approach that brings together environmental and agricultural production, as well as the leisure and recreational functions of rural areas and farming. A similarly multifunctional approach is examined by Kołodziejczak (Chapter 10), who studies agricultural businesses in the European ecological network NATURA 2000 in Poland. While the housing functions of rural areas have been widely recognized in the past (Pacione, 1984), Heffner and Czarnecki (Chapter 11) analyse a special aspect of it, specifically the role of the second home phenomenon in the economic and social development of rural areas in Poland.

An example of rural development outside agricultural production in the global South is given by Mausch and Diez (Chapter 12). They examine the location factors of rural areas for small and medium enterprises (SMEs) in Vietnam and the potential of these SMEs for the creation of non-farm employment as additional income sources.

Finally, Part IV sums up the main findings of these chapters and derives brief general implications for academics, planners, and practitioners in rural areas.

References

Anríquez, G. and Stamoulis, K., 2007. Rural Development and Poverty Reduction: Is Agriculture Still the Key? *Journal of Agricultural and Development Economics* 4, 5–46.

Bathelt, H., 2005. Geographies of Production: Growth Regimes in Spatial Perspective (II) – Knowledge Creation and Growth in Clusters. *Progress in Human Geography* 29, 204–16.

Coe, N., Dicken, P., and Hess, M., 2008. Global Production Networks: Realizing the Potential. *Journal of Economic Geography* 8, 1–25.

D'Hease, L. and Kirsten, J., 2006. *Rural Development – Focussing on Small Scale Agriculture in Southern Africa*. University of Pretoria, Pretoria.

Dannenberg, P., 2009. Die Auswirkungen der NAFTA auf die Landwirtschaft im Integrationsraum. *Geographische Rundschau* 61, 12–18.

———— 2010. Landwirtschaft und ländliche Räume. In: Kulke, E. (ed.), *Wirtschaftsgeographie Deutschlands*. Spektrum, Heidelberg, pp. 75–100.

Dannenberg, P. and Kulke, E., 2005. The Importance of Agrarian Clusters for Rural Areas – Results of Case Studies in Eastern Germany and Western Poland. *Die Erde* 136, 291–309.

Domon, G., 2011. Landscape as Resource: Consequences, Challenges and Opportunities for Rural Development. *Landscape and Urban Planning* 100, 338–40.

Evans, N., Morris, C. and Winter, M., 2002. Conceptualizing Agriculture: A Critique of Post-Productivism as the New Orthodoxy. *Progress in Human Geography* 26, 313–32.

FAO, 2000. Cultivating Our Futures. Taking Stock of the Multifunctional Character of Agriculture and Land.

Gereffi, G., 1996. Global Commodity Chains. New Forms of Coordination and Control among Nations and Firms in International Industries. *Competition and Changes* 4, 427–39.

————— 2001. Shifting Governance Structures in Global Commodity Chains, With Special Reference to the Internet. *American Behavioral Scientist* 44, 1616–37.

Gereffi, G., Humphrey, J., and Sturgeon, T., 2005. The Governance of Global Value Chains. *Review of International Political Economy* 12, 78–104.

Gibbon, P., Bair, J., and Ponte, S., 2008. Governing Global Value Chains: An Introduction. *Economy and Society* 37, 315–38.

Goodman, D. and Watts, M., 1994. Reconfiguring the Rural or Fording the Divide? Capitalist Restructuring and the Global Agro-Food System. *The Journal of Peasant Studies* 22, 1–49.

Henderson, J., Dicken, P., Hess, M., et al., 2002. Global Production Networks and the Analysis of Economic Development. *Review of International Political Economy* 9, 436–64.

Henkel, G., 2004. *Der ländliche Raum*. Gebrueder Borntraeger, Berlin, Stuttgart.

Hughes, D., 2009. European Food Marketing: Understanding Consumer Wants – The Starting Point in Adding Value to Basic Food Products. *Euro Choices – Agri Food and Rural Resource Issues* 8, 6–13.

Ilbery, B. and Bowler, I., 1998. *From Agricultural Productivism to Post-Productivism*. Addison Wesley Longman Ltd.

Isserman, A.M., 2005. In the National Interest: Defining Rural and Urban Correctly in Research and Public Policy. *International Regional Science Review* 28, 465–99.

McCarthy, J., 2005. Rural Geography: Multifunctional Rural Geographies – Reactionary or Radical? *Progress in Human Geography* 29, 773.

Morrison, A. and Rabellotti, R., 2009. Knowledge and Information Networks in an Italian Wine Cluster. *European Planning Studies* 17, 983–1006.

Pacione, M., 1984. *Rural Geography*. Harper and Row.

————— 2009. *Urban Geography – A Global Perspective*. Routledge, London.

Palang, H., Printsmann, A., Gyuro, E.K., et al., 2006. The Forgotten Rural Landscapes of Central and Eastern Europe. *Landscape Ecology* 21, 347–57.

Pollermann, K., Raue, P., and Schnaut, G., 2013. Rural Development Experiences in Germany: Opportunities and Obstacles in Fostering Smart Places through LEADER. *Studies in Agricultural Economics* 115.

Porter, M.E. and Bond, G.C., 1999. *The California Wine Cluster*. Harvard Business School.

PART II
Dynamics in the Food Production Function

Chapter 2

Regional Linkages in the Kenyan Horticultural Industry

Peter Dannenberg and Gilbert Nduru

Introduction

Since the 1980s, exports from Kenya have grown by several hundred per cent and the Kenyan fresh fruit and vegetable (FFV) industry is considered a success story (see below; Barrett et al., 1999; Ouma, 2010).

Newer studies on Kenyan horticulture analysed this success by focusing on integration in the international value chain (Barrett et al., 1999; Dolan and Humphrey, 2000) and on the role of institutions (Jaffee, 1994; Humphrey, 2008). The theoretical framework for this analysis is mostly historical (Minot and Ngigi, 2003), institutional (Jaffee, 1994; Humphrey, 2008), as well as value chain approaches (e.g. Dolan and Humphrey, 2000; Gereffi et al., 2005) with a special focus on the rise of supermarkets and the proliferation of private standards (Graffham et al., 2007; Asfaw et al., 2007; Humphrey, 2008).

This chapter aims to broaden the view of the success factors of Kenyan horticulture by looking at the linkages and networks of the industry itself and within its environments. It is argued that horizontal linkages between farmers and other actors like private services and public extension officers significantly contribute to the competitiveness of the farmers, their bargaining position with the direct buyer and their chances to integrate into the international value chain. This chapter provides an additional explanation for the successful integration of small-scale farmers into international value chains, which has often been neglected in similar studies.

The conceptual background for this chapter includes studies on the success of the Kenyan horticultural system which are linked with work on clusters and creative milieus especially for small-scale businesses (Porter, 1998; Maillat and Lecoq, 1992; Giuliani and Bell, 2005; Dannenberg and Kulke, 2005).

State of the Art on Kenyan Export Horticulture

The Kenyan Horticultural Success Story

Kenya has a strong horticultural tradition in the production of fruits and vegetables (i.e., French beans, mangoes, and snow peas) for export markets (Okado, 2007).

The success and importance of the horticultural industry has increased during the last 25 years after horticulture became a major export industry.

In 2010, the total earnings from horticulture exports reached approximately US$922 million, topping the list of Kenya's largest foreign currency earner as well as being one of the largest suppliers of horticultural products to the European Union (www.trademarksa.org/, 2012). Since it is highly labour-intensive, the industry is a major employer, providing job opportunities both directly and indirectly through associated industries (Okado, 2007).

Although the estimated numbers of farmers involved in horticulture export in Kenya differ significantly, the Kenya Horticultural Development Program (KHDP) estimates that in 2008 about 20,000 farmers (most of them small scale family farmers) grew fresh horticultural products for the export market (see also Ouma, 2010).

Several household surveys showed that farmers producing for export and people employed in farming or related businesses in the industry (i.e., packaging, logistics) are better off than non-export smallholders, earning significantly higher annual household incomes (McCulloch and Ota, 2002; Asfaw et al., 2007; Mwangi, 2008).

The Contemporary Discussion on the Success Factors of Kenyan Horticulture

Some positive factors like the availability of cheap labour, good climatic conditions and general improvements of transportation links to global markets have clearly been outlined (McCulloch and Ota, 2002). However, the poor performance of other countries with similar attributes indicates that Kenyan horticulture possesses advantages that go beyond these obvious ones. Apart from historical factors (in particular the sector's good connections to the market of the former colonial power UK), infrastructural advantages (in particular the Jomo Kenyatta International Airport, the premier East African centre of air transport to Europe), and institutional advantages (in particular a variety of supporting institutions and a liberal legal framework), recent studies especially outline the effective organization of the value chain in Kenya (Asfaw et al., 2007; Dolan and Humphrey, 2000; Ouma, 2010; Dannenberg and Nduru, 2013).

Ouma (2010) as well as Dolan and Humphrey (2000) show that while the EU supermarkets dominate and coordinate the Kenyan horticultural value chain, large portions of the chain's organizational work in Kenya is done by Kenyan exporters, who have both increased their own horticultural production and improved the integration of the Kenyan horticultural industry (Gereffi et al., 2005). Andrew Graffham et al. (2007) and Ouma (2010) underline the role of Kenyan exporters as gatekeepers and supporters (e.g. with training and technical support) of horticultural farmers, especially since the introduction of the private process-orientated EU supermarket standard GlobalGAP, which is the key stipulation for entering the EU market (see also Humphrey, 2008; Mithöfer et al., 2008; Dannenberg and Nduru, 2013). According to Ouma (2010), the requirements of GlobalGAP are so sophisticated that most Kenyan farmers can only achieve them with the support of their exporters. In particular, exporters with high knowledge, technical, and financial *capabilities* help farmers develop integrated 'quality management systems'.

However, Dannenberg and Nduru (2013) showed that still today large numbers of small scale farmers managed to stay integrated in the chain without being supported by exporters. They could partly explain this phenomenon through informal arrangements in which farmers could enter the chain without full formal GlobalGAP certification. However, they did not explain, why large numbers of farmers, including farmers with full formal certification, which did not get exporter support could reach a high level of bargaining power, a high access to business relevant knowledge and a high competiveness which is comparable to the situation of those farmers who are supported by exporters. This chapter will outline this positive situation for these farmers and argue that it can be explained through regional horizontal linkages similar to those which occur in industrial clusters or milieus (Porter, 1998; Maillat and Lecoq, 1992).

A Synthesis of Clusters, Milieus, and Value Chain Approaches

While agricultural economic activities in rural areas of developing countries have been intensively analysed, there are few studies on the development of regional production systems, innovative milieu or regional cluster-like networks in this area. Yet, the positive developments in the Kenyan horticulture industry suggest that such networks have evolved in those regions and contribute to its success. Therefore, this chapter introduces cluster and network approaches (including institutional approaches) and links them to the value chain analyses. This eclectic combination within a synthesis framework allows for a detailed look at regional and interregional linkages. These aspects are analysed in the regional case study for the Mt Kenya horticultural region.

The importance of regional production networks (based on spatial proximity) in the competitiveness of firms was outlined by Alfred Marshall (1920) and regained importance in the 1990s with creative milieus, regional clusters or regional innovation systems (Porter, 1998; Humphrey and Schmitz, 2002).

The fundamental idea behind these concepts is that spatial proximity between companies and supplementary units (i.e., suppliers and institutions) not only leads to classic advantages of agglomerations (i.e., low cost of transport and transactions), but also to the possibility of immaterial exchange relationships (i.e., information, experience; Porter, 1998).

The exchange of information within the network – especially the exchange of non-codified, or partly experience-based knowledge ('tacit knowledge') – can lead to learning processes. Ideally, the resulting 'best practice solutions' for the production process, for improvement of products and innovations increases the competitiveness of the various units within the production system and lead to upgrading processes. Such exchange is mainly possible through personal communication between trusted actors (Dannenberg and Kulke, 2005). As Porter (1998) showed using different empirical examples, in an environment of trust even competitors cooperate ('coopetition') under certain conditions and trust relationships to, for example, improve their buying or selling bargaining power.

Existing case studies on regional agrarian systems (e.g. Dannenberg and Kulke, 2005; Giuliani and Bell, 2005; Porter and Bond, 1999) have outlined some basic elements of these systems: regional agrarian systems can consist of farms, which are supported by networks among each other and by relationships to preliminary units (e.g., suppliers of seeds, fertilizer, production facilities), downstream units (e.g., wholesalers, value-added production, and processing), and to a variety of different service providers (e.g. maintenance, finance, education). An agrarian system can be competitive especially if material input–output relationships are supplemented by immaterial knowledge flows.

As Dannenberg and Kulke (2005) showed using the case study of Poland, regional networks can have positive effects on the competitiveness of small-scale farming systems. Specifically they showed that farmers who were embedded in the regional agrarian systems by mutual formal and informal linkages had better competitiveness (e.g., better bargaining positions) than those who were not. In the case of small-scale farms, joint organization and actions (e.g., knowledge exchange, use of facilities, marketing) were especially important, as these businesses usually did not have the financial and human resources to develop solutions on their own. This is a challenge that most horticultural farmers in Africa also face (Dannenberg and Nduru, 2013).

Nevertheless, as Gereffi et al. (2005) showed, an evaluation of the quality and competitiveness of a regional agrarian system especially in the case of an export based industry must also include a comparative analysis of the networks along the export value chain, which goes beyond the region. Here the existing value chain approaches (Gereffi et al., 2005; Gibbon et al., 2008) are a complementary analytical tool which can be linked with regional production system approaches (Figure 2.1; see also Humphrey and Schmitz, 2002).

As noted earlier the Kenyan production system is dominated by large EU retailers. While there is no agreement about the type of governance (Gereffi et al., 2005), there is a general agreement that the terms and conditions (including the prices, the quality, and the standards under which the products are produced) are set by the large retailers. Accordingly, these retailers provide their supplying actors with the needed information on their requirements (i.e., the GlobalGAP standard).

Under these conditions, it is questionable how far an agrarian system in a rural area might possess influence on the international scale and in the value chain as a whole. However, at the local scale, networks can provide mutual exchange of knowledge (e.g. how to fulfil the standards and best production practices) and improve the farmers bargaining power with direct buyers and suppliers. Additionally, there may even be the possibility that regional networks contribute to a higher degree of independence through the application of joint marketing which might increase the total sales volume. In total, such interlinked systems might lead to higher competitiveness and success of the embedded companies. Figure 2.1 outlines the synthesis between cluster and value chain approaches and shows typical expected linkages.

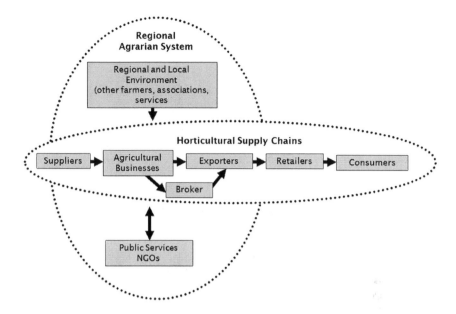

Figure 2.1   Connection between the regional agrarian system and the horticultural supply chain

Source: Own design; see also Dannenberg and Kulke, 2005.

In summary, the following objectives are central to this analysis:

- First, does intensive cross-linking between FFV farms themselves and complementary units (suppliers, buyers, services, and further involved actors) exist at the regional level?
- Second, are the FFV farms involved in the system competitively better off than the non-integrated ones?
- Finally, how much do the different linkages contribute to the success of the FFV farms in relation to each other?

Based on these objectives we draw a conclusion that outlines the relevance of our results in the broader state of the art and for policy applications.

Methods of Data Collection and Analysis

Characteristics of the Case Study Area – Mt Kenya

Because of favourable soils and climatic conditions the Mt Kenya region has one of the largest concentrations of FFV farms in Kenya (Waitathu, 2008).

The number of FFV farms in Kenya and in the Mt Kenya region is volatile, depending on the seasons and differ significantly in studies over the years (see e.g. Ouma, 2010; Mithoefer et al., 2008). According to Ouma (2010: 209), Kenya had about 20,000 farmers growing fresh produce for the export market in 2008. Given a population of 38,610,097 (population census, 2009) and an area of 581,834 km², this leads to a density of 0.52 FFV farms per 1,000 inhabitants and 34.37 farms per 1,000 km² (KNBS, 2012).

For the Mt Kenya region (districts Embu, Kirinyaga, Meru Central, Meru North, Meru South, and Nyeri), Mithofer et al. (2008) came up with roughly 7,200 farms (in 2005) for only four districts (Kirinyaga, Meru Central, Meru North, and Nyeri). While we could not access reliable data for the other two districts 7,200 farms already leads to a concentration of 1.84 farms per 1,000 inhabitants and 530 farms per 1,000 km² for the entire Mt Kenya region,[1] which demonstrates the high structural clustering of export FFV farms in the region (KNBS, 2012).

Most of the farms are individually owned and unaffiliated small-scale operators with less than 10 ha (McCulloch and Ota, 2002). Their horticultural export products include French beans, snow peas, avocados, and mangos (Kaburu et al., 2001; Waitathu, 2008).

Data Collection and Analysis

A methodological blend of expert and stakeholder interviews and standardized questionnaires on the relevant farming businesses was used to gain detailed insight into the networks of the horticultural industry. Overall, a standardized survey involved 169 export-orientated individual FFV farms from 42 randomly selected villages from all different districts of the Mt Kenya region. Five per cent of the surveyed farmers had no formal education; 32 per cent attended only primary school (own results). Additional qualitative interviews were conducted with 32 farmers in the survey and with 42 experts and stakeholders in the region and along the value chain (see Table 2.1).

The experts and institutions were selected based on literature review of existing studies on the industry (e.g. Ouma, 2010; Mithofer et al., 2008). The interviewed brokers were met personally and at random collect points in the region. The importers and exporters were selected randomly via trade directories.

In the survey, the farmers were asked about their immaterial business linkages (business-relevant knowledge exchange and support on fulfilling standard requirements) to various supplementary units. To identify the success of each farming business, the farmers were asked to estimate their annual turnover, bargaining position and future business expectations. The indicators were selected based on interviews with different experts (including Ouma and Kulke). There are a large number of indicators measuring the competitiveness of firms, which all

---

1   At a given population of 3,903,626 people and an area of 13,580 km².

have advantages and weaknesses (Day and Wensley, 1988). Although the two final indicators are based on subjective judgment, when combined with the turnover data and the qualitative results, they give a useful basis for analysing the importance of the business linkages. Furthermore, similar indicators have successfully been used in related studies (Dannenberg and Kulke, 2005).

The data was collected from October 2008 until August 2009. Further interviews for validation took place in October 2012. During the qualitative interviews, the experts and farmers were separately asked to describe and assess the importance of the different linkages and give examples for such interactions (if applicable). Farmers without business-relevant linkages, were asked, why they were not linked. The interviews were manually recorded in order to maintain a trusting and open atmosphere. Afterwards, we analysed and classified the interviews with the social science software Maxqda (which is commonly used for qualitative empirical studies).

The percentage of the quantitative-surveyed farmers who were immaterially linked with the respective actors in the region was outlined to indicate the network activities. Secondly, the farmers were grouped as 'linked' and 'not linked' (knowledge exchange or support) and then compared with competitive performance (indicators: annual turnover, opinions of the farmers on their bargaining position and future expectations) using tabulations.

The quantitative analyses only gave evidence for the existence of direct linkages. Beyond these limits, the qualitative farm interviews indicated the existence of further unquantifiable linkages.

The interpretation of the results and the documentation and explanation of the underlying causalities were completed through quantitative and qualitative comparison of the results and through the interviews with experts, complementary units in the region and other actors along the value chain, and in the context of relevant literature.

Table 2.1   Overview of the interviewees

External experts (scientists)	7
GlobalGAP certifier	3
Retail	5
Importers	8
Exporters	5
Middlemen/brokers	3
Local institutions	11
Farmers	32

Source: Own results.

Results

Vertical Connections in the Industry

Our results show that our sample farmers in the Mt Kenya region are linked to different direct suppliers (i.e. of seeds and pesticides) and buyers through material linkages. In a large number of cases, these linkages go beyond the buying and selling of products, and include knowledge transfer and different ways of support (i.e. in implementing GlobalGAP).

Fifty-nine per cent of the farmers have business knowledge exchanges with their direct suppliers (Table 2.2). This knowledge includes information about how to produce and market their products (i.e., according to the GlobalGAP standard). While this number shows the significance of these linkages, the cross-table calculation shows no positive influence on the farmers' competitiveness (i.e., turnover, bargaining position, or future expectations). The results even indicate a negative correlation. While in European countries, input suppliers often have a close connection to the farmers and train them on the use of their supply (Dannenberg and Kulke, 2005) farmers and suppliers in Kenya stated that suppliers in the area are usually just local traders and only have very limited knowledge. Other farmers directly get their input supply including intensive technical training through their exporters. This positive knowledge exchange was included as 'knowledge exchange with buyers'.

Table 2.2   Overview of vertical linkages of the surveyed farms in the Mt Kenya region

Linkage partners	Percentage (n = 169)	Larger turnover (>100,000 KSH)	Better bargaining position	Better future expectations
Knowledge exchange with supplier	59%	41%	15%	39%
No knowledge exchange with supplier	41%	40%	33%	51%
Total (n = 163)	*100%*	*40%*	*22%*	*44%*
Knowledge exchange with buyer	40%	45%	36%	61%
No knowledge exchange with buyer	60%	38%	12%	33%
Total (n = 166)	*100%*	*41%*	*22%*	*44%*

Source: Own results.

Farmers' competitiveness is highly connected to the linkages with direct buyers. Forty per cent of the farmers stated that they get valuable business knowledge that goes beyond the transaction of goods from their direct buyers. As with the suppliers, this knowledge transfer includes information on how to produce different products, and direct advice on how to fulfil the standards of the international markets. In comparing the 'linked' and 'non-linked' farmers, 36 per cent of the 'linked' farmers saw their bargaining position as at least equal, while only 12 per cent of the non-linked felt the same way. In addition 61 per cent of the linked farmers viewed their future expectations as neutral, good, or very good, while only 33 per cent of the non-linked did. Finally, linked farmers achieved larger turnovers than those who were not linked. This gives first hand evidence that immaterial linkages with buyers improve the situation of farmers. As the buyers differ significantly, it is still necessary to analyse this category further.

The most common way for farmers to sell their products to the export market is through an exporter. This is usually a larger professional company, which manages the transportation of the products with cooled or non-cooled trucks from the farm or a collection point. In this survey 84 per cent of the farmers sold at least part of their products directly to exporters.

The second most common way (54 per cent) is selling to a middleman or a 'broker' (though 38 per cent of farmers sell through exporters and middlemen). In contrast to exporters, these brokers usually have low capabilities. They are mostly located in the region, are much less organized, often possess poor transportation equipment, and are less reliable when it comes to payments and general agreements. Working with brokers is especially common for farmers who are peripherally located and for whom these buyers are often the only connection to export markets. As a farmer from the peripheral northeast of Mt Kenya describes the situation, 'For us it is too difficult to get directly in contact with the exporters because they are so far away' (Farmer A, 15 August 2009). As a result the product marketing of the farmers located in those areas often depends on single brokers, who then have a monopoly of the bargaining position.

Table 2.3   Overview of the distribution channels of the surveyed farms in the Mt Kenya region

Distribution channel	Percentage (n = 169)	Larger turnover (>100,000 KSH)	Better bargaining position	Better future expectations
Exporter	84%	43%	24%	46%
Broker	54%	35%	13%	41%
Total (n = 169)	*100%*	*41%*	*22%*	*44%*

Source: Own results.

This problem also underlines the important function of the brokers as a link to the market for the farmers in peripheral locations. The brokers themselves sell the products either directly to an exporter or sell them to another broker. In some cases, three or more brokers can be involved before the products reach the exporter. Other options like selling to other farmers, or selling to a public institution were only used by a minority (<5 per cent each).

Strong differences are especially evident when comparing bargaining position of the farmers connected to different buyers. Twenty-four per cent of the farmers who sell to exporters state that they have at least an equal bargaining position. Only 13 per cent expressed a similar bargaining position.

A clear difference between the brokers and the exporters lies in their networks. As outlined in several interviews, brokers are usually from a similar background (according to their location, social and cultural ties, and education) as the farmers. Thus, they can only give farmers limited knowledge (see also Dannenberg and Kulke, 2005) which is often 'second hand' such as on international market requirements:

> If you sell to broker you do not know so much about GlobalGAP, they often only noted what the neighbours do. (Exporter A, 10 August 2009)

In comparison, the networks of the exporters have four advantages:

- Most exporters have established quality management systems for farmers that allow the exporters to supervise and advise their farmers on management and production methods through technical assistants on the farm (Jaffee, 1994; Minot and Ngigi, 2003).
- Most exporters are located in Nairobi and are therefore more closely linked to the government offices, businesses, and donors that shape horticultural policies and other key developments. Therefore, they have and can offer quick and first-hand information (personal interviews with Exporter A, 16 August 2009; and Exporter B, 19 August 2009).
- In their role as exporters they are also directly linked to their international counter parts, which give them quick information on ongoing developments in the markets.
- In addition, most exporters belong to the Fresh Produce Exporters Association of Kenya (FPEAK). While FPEAK generally supports its members with information, lobbying and marketing, one of the key functions of FPEAK is also the promotion of members' compliance with international standards. As the CEO of FPEAK stated, 'We are monitoring Japan, EU and US standards [...] and we then advise our members as soon as possible, so that they can comply on time' (Exporter C, 10 August 2009). Because of the support of FPEAK, the exporters teach their suppliers how to fulfil required market standards (FPEAK, 2008).

Some of the exporters are also producers with large capital intensive production units in the region. However, these exporter-owned production units usually do not directly compete with the smaller individual farmers, but work in close cooperation with them to increase joint export sales volumes. While the number of these exporters was too small for meaningful quantitative analysis, the interviews suggested that the linkages to these larger companies had especially positive effects on the competitiveness of both larger and smaller farms. As the manager of a large scale exporting and producing company described:

> We need the small farmers and they need us. We help them with training and information and even let them work on our farms but we also buy their products during peak demand, when we do not have enough volume to sell. However, we only can buy their products if they fulfil the requirements of our buyers, so that is why we train them. (Large-Scale Farmer C, 14 August 2009)

In this way, large-scale and small-scale farmers are not only acting as competitors but also as partners in certain areas of joint production and knowledge exchange (coopetition; Porter, 1998).

This leads to the conclusion that interregional and internationally based exporters, with their historically developed quality management systems and their good international connections, can be seen as key buyers with access to key knowledge about the market. While Peter Gibbon et al. (2008) use the term 'turn key supplier' as crucial interfaces to organize the flows in the value chain from a perspective at the end of the chain, this case shows the perspective from the farm gate.

While the importance of the linkages to the exporters has already been highlighted, it is interesting, that large numbers of farmers are not directly linked to exporters, but still sell through brokers, who usually do not have good access to relevant business information. The question arises as to whether there are other networks or sources, which may support the small-scale farms to produce and fulfil the high requirements of EU buyers.

Horizontal Business Connections

Regarding the linkages between farmers themselves, the majority of the farmers (85 per cent) are members of local or regional farmers associations that are formally organized and usually registered with the Kenyan Social Service department or the Ministry of Cooperative Development and Marketing. The associations hold meetings and assemblies where they elect their leaders and discuss their business problems. Through these associations, the farmers can access training (i.e. private consultants) on various aspects of horticultural production (i.e., pruning, fertilization) and receive help meet export standards such as GlobalGAP.

Table 2.4   Overview of horizontal linkages of the surveyed farms in the Mt Kenya region

Linkage partners	Percentage (n = 169)	Larger turnover (>100,000 KSH)	Better bargaining position	Better future expectations
Member in an association	85%	43%	25%	44%
No member in an association	15%	28%	8%	48%
Total (n = 169)	*100%*	*41%*	*22%*	*45%*
Knowledge exchange with cooperatives	9%	67%	27%	60%
No knowledge exchange with cooperatives	91%	38%	22%	42%
Total (n = 163)	*100%*	*40%*	*22%*	*44%*
Knowledge exchange with private consultants	7%	64%	36%	73%
No knowledge exchange with consultants	93%	39%	21%	41%
Total (n = 163)	*100%*	*40%*	*22%*	*44%*

Source: Own results.

The survey showed positive correlations between membership of a local or regional association and higher FFV success in the turnover and bargaining positions, although, there is no clear correlation between being an association member and the future expectations of the member. Also, our qualitative interviews revealed membership advantages for association members:

- The exchange of knowledge about different buyers and their reliability.
- The pooling of their produce to achieve volumes that make them attractive to more buyers and give them better market options.
- The creation of direct connections with exporters, which leaves out middlemen and provides exporters with single negotiation partners. One farmer stated: 'When we organized ourselves we could go directly to the exporter and we had a better bargaining position for the prices. With our organization we could also check out different exporters in

Nairobi and bargained. So it was a self-empowerment' (Farmer D, 10 September 2009).
- The ability to pressure buyers to deal with them fairly or lose business with the entire group.

Apart from official associations, the majority of the farmers were organized in their local villages (mostly embedded into village hierarchies led by village elders).[2] In these communities, the cooperation varied but often included collective storage of the products (i.e. in a joint shed), buying farm supplies in order to lower costs, and jointly negotiating with buyers. A chief of a group of 20 farmers organized at the village level described their cooperation: 'I as a chief organize and buy the supply for the whole group. I also get in contact with our buyer and we exchange information [...] In the past, they have shown us how to produce but also how to build storages' (Farmer E, 15 August 2009). This statement also underlines the importance of the group in improving linkages with external actors. Another aspect of group actions in several villages is the building of joint charcoal-powered cold storage for perishable horticultural produce. While this form of cooperation was commonly practiced informally (and could therefore not be recorded in the questionnaire), 9 per cent (in total 15 farmers) were members of a formal cooperative. While this number is low (and therefore quantitative evidence is limited), our survey underlines the positive effects of cooperation as a cooperative in all three indictor variables.

A third horizontal linkage that is positively discussed in cluster approaches is the linkage that farmers have with private consultants. In our sample, private services (i.e. banks, consultancies and trainers) were generally used by larger companies and organized groups to share the relatively high costs of such services. The survey indicated that only 12 farmers individually had business relevant immaterial linkages.[3] Group trainings by professional trainers were seen as critical sources of knowledge for all 42 villages that we visited. Most of these training sessions focused on proper usage of chemicals, storage solutions for perishable goods, and the fulfilment of standards.

Local banks offer farmers additional support and training. For example, Equity Bank employs 25 horticultural specialists in the Mt Kenya region. These specialists conduct free training sessions that are open to entire villages at local farms. In these training sessions, specialists e.g. demonstrate production methods. As one of the officials noted: 'The idea behind it, is of course, that those farmers we help will go to us if they need credits, and if we help them running a successful business, they become better customers' (Officer of the Equity Bank, 11 August 2009).

---

2   The concrete number is hard to tell as the level of organization goes hand in hand with the general organization of the village and is mostly informal.

3   Those immaterial linkages transmit business information (especially the exchange of knowledge on production and marketing) that goes beyond normal services like banking.

Connections with Institutions and the Further Local Environment

Apart from private business relations, different support units could be identified in the local and regional environment.

Most farmers stated that they had good relations with the local municipalities.[4] Based on interviews with the farmers, interaction with public extension officers varied strongly according to the quality and the availability of the public extension officers. In general, the officers aimed to help and consult farmers in different areas of production and marketing. Some local districts had a variety of officers with specialties on certain crops, while others just had one or two officers, with low qualifications. There were clear differences in the quantity and intensity of the linkages depending also on the infrastructural connection of the villages.

Apart from the local governmental support at the municipality level, our study could identify the Horticultural Crops Development Authority (HCDA) as a supporting unit in the Mt Kenya region. While the general supportive effects of the HCDA have already been outlined in different studies (see above) its activities at the regional level have less of a positive impact. In the Mt Kenya region, HCDA has several projects, including the building of large cooling stores for fresh fruit and vegetables and other forms of logistical support. Unfortunately, as both farmers and officers of HCDA stated these projects have so far not led to larger benefits, because there is no agreement between farmers and HCDA as to who is responsible for the costs of running the facilities. One officer in charge of an HCDA depot described the current situation clearly: 'Our facilities are too expensive for most farmers and are not used much' (Institution A, 13 August 2009). In contrast, the transport, cooling, and packing facilities at Jomo Kenyatta International Airport are highly frequented.

As mentioned above, another group of supportive actors in horticultural businesses, which have been controversially discussed in literature (see e.g. Humphrey, 2008) are the different international and national donors. According to our study results, the role of donors seems to be rather marginal. There is generally a large variety of national and international donors in Kenya. Even so, only 5 per cent of the farmers stated that they got any donor support. This is partly because donor connections to the farmers in Mt Kenya villages are limited by bad roads (different interviews with farmers; see also Dannenberg et al., 2011). As an officer of an international donor outlined, 'Of course the main parts of our activities take place in the rural areas, but it is impossible to reach everybody' (Institution B, 18 August 2009). Another reason for the limited success of the donor activities is the lack of trust. As one farmer stated: 'farmers are not cooperative with foreigners, they do not accept foreign aid. They need trustful people' (Farmer F, 12 August 2009). This was underlined by an expert from Moi University: 'The farmers have much more trust in local organizations' (Expert A, 17 August 2008).

---

4   It was not possible to quantify these networks, as the contacts to municipalities, public extension officers and donors take mainly place at the village level and not directly with the farmer.

Table 2.5 **Overview of the linkages with institutions and the local environment of the surveyed farms in the Mt Kenya region**

Linkage partners	Percentage (n = 169)	Larger turnover (>100,000 KSH)	Better bargaining position	Better future expectations
Exchange with local environment	40%	52%	23%	43%
No exchange with local environment	60%	33%	21%	44%
Total (n = 163)	*100%*	*41%*	*22%*	*45%*
Exchange with family	21%	56%	12%	53%
No exchange with family	79%	36%	25%	41%
Total (n = 163)	*100%*	*40%*	*22%*	*44%*

Source: Own results.

While the analysed actors all came from a professional farming-related background, the study also identified that 40 per cent of the farmers also used private local contacts (i.e. family members, friends, and neighbours) as sources of knowledge exchange and support. The most important source of this was family, with 21 per cent (54 per cent of all named sources).

The impact of these linkages on the competitiveness of the farms is inconclusive. Our interviews reveal a lack of competence of the actors in the local environment in specific aspects of farming, especially where qualified knowledge is needed such as for meeting international standards. Here, local friends and family networks could not substitute professional help (interviews with various farmers and experts, 2008–2009).

Conclusion and Outlook

The results show the existence of different horizontal and vertical networks in the Mt Kenya region, which go beyond input–output, command, and control relations in the value chain. Buyers and suppliers are seen and are frequently used as business relevant knowledge sources by large parts of the farmers. Apart from these vertical linkages a variety of horizontal business relationships and connections with institutions and the local environment exist. However, these linkages vary strongly according to their total number and their impact on the competitiveness of the farmers.

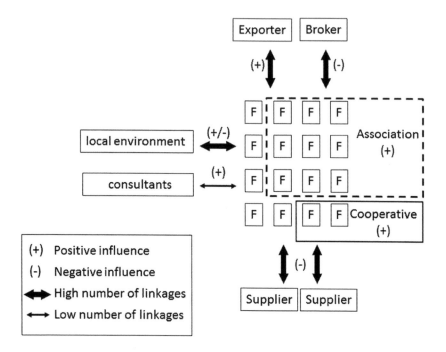

Figure 2.2   Key knowledge and support networks in export-orientated FFV farming in Mt Kenya

Source: Own design.

The example of the relation between small-scale and large companies reveals that this structural mixture leads to synergies for both farming groups. Valuable knowledge spillovers even take place between trusted local competitors ('coopetition'). While the results outline the key role of exporters, the large number of competitive producers not working directly with exporters but linked with regional actors and institutions shows that integration in regional networks positively affects the competitiveness of the embedded farms. The comparison of 'linked' and 'non-linked' farmers shows that it is not only the structure (i.e., the existence of valuable organizations and institutions) of the regional agrarian system (or a complete industry) that matters, but also the interaction that take place within the system. This study emphasizes the importance of local networks for less developed countries and regions with poor transport systems and limited access to information. However the study also shows that the different linkages also differ in their impact on the competitiveness. For example, the knowledge exchange with (typically low-qualified) suppliers and family members has not improved the competitiveness of the farmers.

In terms of the discussion of global value chains, our results suggest that a focus on value chains, power relations, and institutional frameworks alone, without considering the importance of the linkages, networks, and structures in regional agrarian systems leaves out important explanations for the success of the Kenyan horticultural industry.

Our study can also give new additional knowledge insights for developing and improving regional policy and donor support systems. This could include training measures for the suppliers and brokers who have high numbers of contacts with farmers and could therefore be used as multipliers in areas where accessibility for donors is difficult and where farmers are not integrated in exporter-based quality management systems.

While the study so far indicates, that horizontal linkages occur mainly in local and regional networks so far, future research should also consider possible broader linkages with actors outside the region, which might develop due to an increasing use of information and communication technologies (ICT) for businesses (Dannenberg and Lakes, 2013).

References

Asfaw, S., Mithöfer, D., and Waibel, H., 2007. What Impact Are EU Supermarket Standards Having on Developing Countries Export of High-Value Horticultural Products? Evidence from Kenya. In: 105th EAAE Seminar, 'International Marketing and International Trade of Quality Food Products', Bologna.

Barrett, H., Ilbery, B., Browne, A., et al., 1999. Globalization and the Changing Networks of Food Supply: the Importation of Fresh Horticultural Produce from Kenya into the UK. *Transactions of the Institute of British Geographers* 24 (2), 159–74.

Dannenberg, P. and Kulke, E., 2005. The Importance of Agrarian Clusters for Rural Areas – Results of Case Studies in Eastern Germany and Western Poland. *Die Erde* 136 (3), 291–309.

Dannenberg, P., Kunze, M., and Nduru, G., 2011. Isochronal Map of Fresh Fruits and Vegetable Transportation from the Mt Kenya Region to Nairobi. *Journal of Maps* 11 (0), 273–9.

Dannenberg, P. and Lakes, T., 2013. The Use of Mobile Phones by Kenyan Export-Orientated Small-Scale Farmers – Insights from Fruit and Vegetable Farming in the Mt Kenya Region. *Economia Agro-Alimentare* 2013 (3), 55–76.

Dannenberg, P. and Nduru, G., 2013. New Challenges and Realities in International Value Chains – Analysing the Proliferation of Standards beyond the Exclusion Debate. *Tijdschrift voor Economische en Sociale Geographie* 104 (1), 41–56.

Day, G. and Wensley, R., 1988. Assessing Advantage: A Framework for Diagnosing Competitive Superiority. *Journal of Marketing* 52 (2), 1–20.

Dolan, C. and Humphrey, J., 2000. Governance and Trade in Fresh Vegetables: The Impact of UK Supermarkets on the African Horticulture Industry. *Journal of Development Studies* 37 (2), 147–76.

FPEAK, 2008. Fresh Produce Exporters Association of Kenya [online]. Available at: http://www.fpeak.org/ [accessed: 15 November 2013].

Gereffi, G., Humphrey, J., and Sturgeon, T., 2005. The Governance of Global Value Chains. *Review of International Political Economy* 12 (1), 78–104.

Gibbon, P., Bair, J., and Ponte, S., 2008. Governing Global Value Chains: An Introduction. *Economy and Society* 37 (3), 315–38.

Giuliani, E. and Bell, M., 2005. The Micro-Determinants of Meso-Level Learning and Innovation: Evidence from a Chilean Wine Cluster. *Research Policy* 34 (1), 47–68.

Graffham, A., Karehu, E., and MacGregor, J., 2007. *Impact of EurepGAP on Small-Scale Vegetable Growers in Kenya*. Fresh Insights (Cmnd. 6), London, International Institute for Environment and Development.

Humphrey, J., 2008. Private Standards, Small Farmers and Donor Policy: EUREPGAP in Kenya. IDS working paper (Cmnd. 308), Brighton, Institute of Development Studies.

Humphrey, J. and Schmitz, H., 2002. How Does Insertion in Global Value Chains Affect Upgrading in Industrial Clusters? *Regional Studies* 36 (9), 1017–27.

Jaffee, S., 1994. Contract Farming and Shadow Competitive Markets: The Experience of Kenyan Horticulture. In: Watts, M. and Little, P. (eds), *Living Under Contract. Contact Farming and Agrarian Transformation in Sub-Saharan Africa*. Madison, University of Wisconsin Press, pp. 97–139.

Kaburu, K., Kangara, J., Ngoroi, E., et al., 2001. *Diversity of Vegetables and Fruits and Their Utilization Among the Nduuri Community of Embu, Kenya* [online]. Available at: http://archive.unu.edu/env/plec/clusters/Eastafrica/nov2001/Kaburu.pdf [accessed: 15 November 2013].

Kenya National Bureau of Statistics, 2012 [online]. Available at: http://www.knbs.or.ke [accessed: 15 November 2013].

KHDP, 2007. Kenya GAP Launched. Update on Kenyan Horticulture. Nairobi: KHDP.

Marshall, A., 1920. *Industry and Trade*. 8th Edition. London, Macmillan.

McCulloch, N. and Ota, M., 2002. Export Horticulture and Poverty in Kenya. IDS working paper (Cmnd. 174), Brighton, Institute of Development Studies.

Minot, N. and Ngigi, M., 2003. *Are Horticultural Exports a Replicable Success Story? Evidence from Kenya and Côte d'Ivoire*. InWEnt, IFPRI, NEPAD, CTA conference, 'Successes in African Agriculture', Pretoria, 1–3 December 2003.

Mithofer, D., Nang'ole, E., and Asfaw, S., 2008. Smallholder Access to the Export Market: The Case of Vegetables in Kenya. *Outlook on Agriculture* 37 (3), 203–11.

Mwangi, T., 2008. Impact of Private Agrifood Standards on Smallholder Incomes in Kenya. Kenya Horticultural Development Program [online]. Available at: http://www.agrifoodstandards.net/en/filemanager/active?fid=135 [accessed: 15 November 2013].

Okado, M. 2007. Lessons of Experience From the Kenyan Horticultural Industry. United Nations Conference on Trade and Development, Nairobi, 29–31 May 2001.

Ouma, S., 2010. Global Standards, Local Realities: Private Agrifood Governance and the Restructuring of the Kenyan Horticulture Industry. *Economic Geography* 86 (2), 197–222.

Porter, M., 1998. Location, Clusters and the 'New' Microeconomics of Competition. *Business Economics* 33 (1), 7–17.

Porter, M.E. and Bond, G.C., 1999. *The California Wine Cluster*. 1st Edition. Harvard, Harvard Business School.

Trademarksa, 2012. Kenya Food Exports Underpinned by Rising Prices [online]. Available at: http://www.trademarksa.org/news/kenya-food-exports-underpinned-rising-prices [accessed: 15 November 2013].

Waitathu, N., 2008. Kenyan Horticulture: Weathering the Political Storm? [online]. Available at: http://www.new-agri.co.uk/08/02/develop/dev1.php [accessed: 15 November 2013].

Chapter 3

Factors Influencing the Market Linkage of Organic and Conventional Tomato Farming Systems in Karnataka, India

Nithya Vishwanath Gowdru,[1] Wolfgang Bokelmann,[2]
Ravi Nandi,[3] and Heide Hoffmann[4]

Introduction

Globally, around 1.5 billion people are engaged in smallholder agriculture and they make up 75 per cent of the world's poorest people whose food, income, and livelihood prospects depend on agriculture (Ferris et al., 2014). Smallholder farmers are already facing numerous risks to their agricultural production which often undermine their household food and income security (O'Brien et al., 2004; Morton, 2007). Since smallholder farmers typically depend on agriculture for their livelihoods and have limited resources and capacity to cope with shocks; any reduction to agricultural incomes can have significant impacts on their food security, nutrition, and wellbeing (McDowell and Hess, 2012). In recent years, there is growing concern of the risks and vulnerability among low-income households in developing countries. Traditionally, risk and vulnerability in farming were addressed with the help of crop diversification, crop insurance (Siegel and Alwang, 1999), enterprise diversification (Siegel and Alwang, 1999) participation in self-help groups (SHGs),[5] and contributing to savings (Puhazhendi and Badatya, 2002; Armendáriz and Morduch, 2010). Presently, climate change and variability are a considerable threat to agricultural communities; threats include extreme weather

1   nithyavishwanath@gmail.com, Humboldt-Universität zu Berlin, Landwirtschaftlich-Gärtnerische Fakultät Ökonomie der Gärtnerischen Produktion, Invalidenstrasse 42, 10115 Berlin, Germany.

2   w.bokelmann@agrar.hu-berlin.de, Humboldt-Universität zu Berlin, Land-wirtschaftlich-Gärtnerische Fakultät Ökonomie der Gärtnerischen Produktion, Invaliden-strasse 42, 10115 Berlin, Germany.

3   nandi999hu@gmail.com, Humboldt-Universität zu Berlin, Landwirtschaftlich-Gärtnerische Fakultät Ökonomie der Gärtnerischen Produktion, Invalidenstrasse 42, 10115 Berlin, Germany.

4   heide.hoffmann@agrar.hu-berlin.de.

5   SHGs: self-help groups are a social design in which people participate by making themselves socially and economically accountable to each other.

conditions, increased water stress and drought, and desertification, as well as adverse health effects (Müller, 2009). If society fails in adaptation then adverse effects are likely to multiply and overstretch many societies' adaptive capacities, which may lead to destabilization and security risks, including malnutrition, loss of livelihoods, forced migration, and conflict (Rice, 2007; Lobell et al., 2008). The Bali Action Plan from the United Nations (UN) climate change conference in Bali in 2007 (Aldy and Stavins, 2007) clearly emphasizes the importance of enhanced action on adaptation. However, vulnerability to climate change can be exacerbated by the presence of other stresses, that 'future vulnerability depends not only on climate change but also on development pathways', and that 'sustainable development can reduce vulnerability to climate change, and climate change could impede nations' abilities to achieve sustainable development pathways' (Rice, 2007).

The nature of risks in agriculture is also changing substantially. In recent times, organic farming has increasingly gained attention as a way to manage natural resources in a more sustainable way and to raise incomes, especially, for smallholder farmers in developing countries (Eyhorn, Ramakrishnan, and Mäder, 2007). In addition, organic farming has experienced a considerable rise in most of the industrialized countries over the past decades, where, developing countries were involved in the organic market mainly as suppliers (Eyhorn, 2007). Especially in developing countries, organic farming adoption rate is increasing (Hine and Pretty, 2006). But this later developments haven't meant an automatic conversion in rural income. In this sense markets, market access and product price are critical for improving rural income. One of the reasons for the poor product prices is the role of superfluous middleman in developing countries (Quartey et al., 2012), like India, which has interlocked markets that preclude farmers from selling their produce by catching market signals of better prices elsewhere. In addition, lack of market linkages add to the predicament especially in the case of perishable produce, which is more common in horticulture crops. However, with the changing global agricultural economy, rise in commodity prices and complex value chains, there is a growing interest in how farmers can benefit from emerging market opportunities (Hellin, Lundy, and Meijer, 2009).

In the case of India, the strategy to seek premium prices for organic produce through quality, food safety, certifications, and labelling has been adopted by many producers as well as promoted by public and private institutions. However, at a global level, the market for these products is more imperfect than the conventional ones (Jaeck, Lifran, and Stahn, 2012). Efficiency losses caused by market failures have been one of the most important drivers to encourage producers' associations and other linkages across the market (Casellas, Berges, and Calá, 2006). Further, a growing recognition for smallholder farmers to shift from conventional farming to innovative organic farming has led to better farm incomes. One such example of innovative, organic farming relates to the establishment of linkages to the producers and markets (FAO, 2008).

Particularly in India, participation of smallholder farmers in domestic markets remains low due to a range of constraints. One of the major constraints faced by smallholder farmers is linked to poor market access. Market access is important because it acts as a mechanism for exchange and helps derive benefits such as income and open opportunities for rural employment (market activities: processing, transportation, and selling create employment); involvement of smallholder farmers in markets contributes to poverty alleviation and is important for sustainable agriculture and economic growth (Guidi, 2011). For farmers, growing and harvesting a crop and rearing animals form only half of the battle as they still have to market the produce. With this background, the purpose of this study was to identify farm level factors that influence smallholder market participation.

Methodology

Study Area

The study was conducted in Karnataka, India, the pioneer state in the country to implement an organic farming policy during 2004 for the promotion of organic farming. The state is known for its rich biodiversity and bestowed with 10 agro climatic zones. It is the ninth largest populated state in the country with 61.09 million people (census, 2011), of which 61.33 per cent reside in rural areas and 73 per cent depend on agriculture. In 2011, 1.56 million ha was devoted to horticultural crop production; out of which, 0.33 million ha was allocated for vegetables production. Among all vegetables, tomatoes stand first with annual production at 952,849 MT within 35,429 ha (census, 2011). The state is divided into 27 districts, out of which, Bangalore rural and Kolar districts were selected for the study because of their extensive cultivation of tomato farming around the year (NHB, 2012). These two districts are located on the Deccan Plateau in the south-eastern part of Karnataka (12.97° N 77.56° E). Bangalore rural district (2,260 km^2) has 0.98 million inhabitants, of which 77 per cent are engaged in agriculture. The area under agriculture is 65 per cent of the total area. The majority of the farmers are small-scale and marginal. The district is well-known for its fruit and vegetable production in the state; among the vegetables, tomato is the major crop cultivated in the district. Further, Kolar district has a population of 1.54 million in an area of 3,969 km^2, of which 0.19 million ha is under agricultural cultivation. Similar to Bangalore rural district, Kolar also has small-scale and marginal fruit and vegetable growers. The district stands first for its tomato production in the state (KSDA, 2014).[6]

Rural Bangalore and Kolar districts were selected by the State Department of Agriculture for the project Organic Adoption and Certification, under the Karnataka State Horticulture Mission Agency (KSHMA), with an aim to convert

---

6   KSDA: Karnataka State Department of Agriculture, http://raitamitra.kar.nic.in/ENG/index.asp.

850 ha of land to organic farming, through farmer cluster development, training, internal control system (ICS), certification, and market-linkages. Considering the above-mentioned facts and after discussions with experts on their opinion of the potential of organic farming in the districts, these two districts were selected as the study area.

Selection and Description of the Sample

The source of information used in this study was mainly obtained from household heads through face-to-face interviews. In the absence of the head, the spouse or any family member directly involved in the farming activities and management was interviewed based on structured questionnaires. In total, a sample of 50 organic and 50 conventional smallholders were interviewed. A purposive random sampling was drawn from an official list of certified and non-certified organic farmers from the International Competence Center for Organic Agriculture (ICCOA) and the state department of agriculture of Karnataka state. Data was gathered from the survey from November 2011 to February 2012. The validity of the questionnaires was assessed by a panel of experts from the Department of Agriculture, experts from a non-governmental organization (NGO), and industry experts in the state. As of 2011, these farmers had at least three years' experience of certified organic tomato cultivation. The interviews included both open ended and closed questions, some of which also elicited quantitative data. For conducting the final interview, the questionnaire was designed through extensive pre-testing of each particular question via personal interviews with a few farmers. Bearing in mind the aims of this study, the farmers selected were smallholders who have less than two hectares of agricultural land and cultivate vegetables. The data was analysed with the Statistical Package for the Social Sciences (SPSS).

Statistical Analysis

Firstly, a logistic regression model was chosen because of its ability to determine the effect of variables on the probability of market linkages. Secondly, it yields the highest predictive accuracy possible with a given set of predictors (Aldrich and Nelson, 1984). Logit analysis was estimated to find the probability (P_i) that farmers have market linkage, as influenced by factors influencing market linkage. The dependent variable is dichotomous, taking two values: 1 for farmers who have market linkage, 0 otherwise. X_i is the independent variable determining Y, shown in Table 3.1.

The relationship between P_i and X_i is not linear. The probability that farmers would cultivate an organic crop approaches 0 at a slower and slower rate, as X_i becomes small, and the probability approaches 1 at a slower and slower rate as X_i becomes large. Since, $P_i = E(Y_i) = 1$ given X_i, non-linearly increases with X_i, P_i is a logistic function of Z_i, given by:

$$P_i = \frac{1}{[1+e^{(-Z_i)}]}$$

Where $Z = A + \sum B_i X_i$

As Z ranges from $-\infty$ to $+\infty$, P_i ranges from 0 to 1, and that P_i is non-linearly related to Z. If

$$P_i = \frac{1}{[1 + e^{(-Z_i)}]} \quad \text{------------ (1)}$$

subtracting P_i from 1 on both sides in equation (1), we have,

$$1 - P_i = 1 - \frac{1}{[1 + e^{(-Z_i)}]}$$

or,

$$1 - P_i = \frac{e^{(-Z)}}{[1 + e^{(-Z_i)}]}$$

$$1 - P_i = \frac{1/e^{(Z)}}{1 + 1/e^{(Z)}}$$

then,

$$1 - P_i = \frac{1}{1 + e^{(Z_i)}} \quad \text{------------ (2)}$$

Therefore, from equations (1) and (2),

$$\left[P_i / (1 - P_i) \right] = e^Z$$

Here $[P_i/(1-P_i)]$ is called the odds ratio, which indicates the ratio of the number of chances in favour of the farmer's willingness to cultivate organic crops to the number of chances of not cultivating crops. Taking the logarithm of this odds ratio to the base e, we get,

$$\text{Log} \left[P_i / (1-P_i) \right] = Z = A + \sum B_i X_i \quad \text{------------} \quad (3)$$

$$L^* = Z = A + \sum B_i X_i \quad \text{------------} \quad (4)$$

Here, L* is the logit as it follows logistical distribution.

For better interpretation of β coefficients, the antilog of β is calculated and the function takes on the following form:

$$\left[P_i / (1-P_i) \right] = e^{\beta 1 + \beta 2 X_i + ui} \quad \text{------------} \quad (5)$$

Table 3.1   Variables and units of measurements of key variables modelled

Variables	Definition
Dependent variable Y	1 if farmer has a market linkage, 0 otherwise
Independent variables	
Age (AG)	Number of years of age of the household
Education level (EDU)	Illiterate, 1; primary, 2; high school, 3; college, 4
Family labour size (FLBR)	Number of family members working on the farm
Dairy activities (DA)	Activities undertaken other than farming, 1 = yes, 0 = no
Contact with extension agent (EXTNCONT)	Number of times per year
Participation in training activities (TRAG)	Participation in training and visits (times per year)
Getting market information through mobile phones (ICT)	1 = yes, 0 = no
Farm size (FRMSZ)	Ownership of the farm in acres
Quality test and certification for their produce (CERT)	1 = yes, 0 = no

Results and Discussion

Both industry and academic studies have investigated the socio-demographic profiles of organic farmers across the world, and to date, these studies have yielded conflicting results. In this study, the socio-demographic profiles of the smallholder of both organic and conventional farmers were analysed and results reveal that (Table 3.2), the average age of organic farmers is 37.52 years whereas the conventional farmer's average is higher – 45.62 years. According to Singh and George (2012), the average age of the conventional farmers in India is 43 years, which shows that the organic growers are younger than conventional growers. Conversely, as reported by Bourn and Prescott (2002) and Rezvanfar, Eraktan, and Olhan (2011), the average age of conventional farmers in Iran and Thailand are 39.5 and 43.74 years respectively, as against 40 and 48.6 years respectively in the same study for organic farmers, which means that, the organic farmers are older than the counterpart. The education level among the farmers sampled is generally low. However, the organic farmers had a higher education compared to conventional farmers. These results are in line with other studies which state relatively high levels of literacy and education of organic farmers mainly when compared to conventional farmers (Duram, 1999; Iliopoulou, Douma, and Giourga, 2011; Chouichom and Yamao, 2010). Furthermore, organic farmers were more affiliated with different institutions, with frequent extension contact and were involved in different training activities compared to conventional farmers in the study area. These results, in agreement with correlate studies (Fairweather, 1999), reveal that institutional support is one of the important motivating factors for the adoption of organic farming among small-scale farmers.

Despite the continuous efforts made by various agencies to promote organic farming, official records have shown that as of 2013, the total area under organic farming in Karnataka was 0.08 mha, which is negligible compared to the total area under agriculture 12.16 mha (KSDA, 2014).[7] However, the result of the field survey conducted as a part of this study revealed that the conventional farmers had a larger farm size compared to organic farmers. This indicates that organic vegetable cultivation was practiced mainly by smallholders. These results are in line with the study conducted by Thapa and Rattanasuteerakul (2011) in Thailand. Further, these results contrast with the reviews of comparative data on organic farming. The latter highlighted that the average farm holding size of organic farms are larger than conventional ones (Offermann and Nieberg, 2000; Chouichom and Yamao, 2010). Even though the organic farmers had less land, the productivity is higher with less family labour compared to conventional farmers in the study area.

---

7   KSDA: Karnataka State Department of Agriculture, http://raitamitra.kar.nic.in/ENG/index.asp.

Table 3.2   Descriptive statistics of organic and conventional farmers

Variables	Mean	
	Organic farming	**Conventional farming**
Age	37.52	45.62
Education	2.76	2.04
Family labour	3.72	3.26
Allied activities	0.84	0.32
High productivity	3.84	2.16
Extension contact	3.52	1.42
Training activities	2.9	0.7
Social network	3.76	2.3
Affiliation with institutions	1	0.52
Farm size	4.063	5.88

Source: Author's own compilation.

In most developing countries, the smallholder farmers tend not to be organized in the markets as they usually sell their limited agricultural produce surplus individually and directly to traders or consumers without linking to the market. In other words, smallholder farmers lack collective action in markets. Individual marketing of small quantities of produce weakens the smallholder farmers' bargaining positions and often exposes them to price exploitation by traders. They also do not benefit from economies of scale. In the study area, the majority of the organic farmers interviewed were practicing collective action in production and marketing their produce where the collective action enables them to deal with transportation, storage issues, acquire technology, get group certification for their produce to comply with the required quality standards and also to supply the desired quantity of their produce. We can also find similar kinds of examples from other developing countries, for example Markelova et al. (2009) did case studies on the role of collective action institutions in improving market access for the rural poor and also examined what conditions facilitate effective producer organizations for smallholders' market access in Kenya and Tanzania. Results reveal that collective action for marketing purposes enable smallholders to overcome multiple market imperfection in the developing world and deal with the high transaction costs associated with the marketing.

The factors which influence market linkage are presented in Table 3.3. The results of coefficients in a logistic regression model represent the amount the logit of the probability of the outcome changes with a unit increase in the predictor. The result reveals that the farmers with dairy activities have a 2.73 times higher chance of having market linkage than non-dairy farmers. Animal products are an integrated part of a holistic agriculture and important for nutrient recycling and other factors. In organic agriculture, cattle are necessary for the farm, as they help to maintain the fertility of the farm, diversify production, spread risk, and increase income (Odhong et al., 2014). Further, the result shows that when dairy activities are not undertaken by the farmers in the study area then the probability of having a market linkage is only 37 per cent. A group of authors (Chander et al., 2011) reported that organic livestock production is an emerging opportunity for producers in developing countries. In addition, well-managed organic dairy farms can reduce many of the environmental and public health risks associated with most conventional dairy farms (O'Hara and Parsons, 2012). Nowadays, consumers are also well-educated and demand high quality for the goods they buy. In addition, they will not buy food products unless there is a guarantee that they are safe to consume. In other words, arguably, consumers make purchasing decisions depending on packaging, consistency, as well as uniformity of goods. Most smallholder farmer's crops have no clearly defined grades and standards and therefore, cannot meet consumers' demands. The reason is that the farmers lack the knowledge and resources to ascertain such requirements. In addition, institutions for determining market standards and grades tend to be poorly developed in the smallholder farmer's environment. Due to uncertainty about the reliability and quality of their goods, they usually cannot get contracts to supply formal intermediaries such as shops and processors. The present study results reveal that farmers with expertise in grades and standards and organic certification have a 1.88 times greater chance of having market linkage than non-certified farmers. This indicates that only well-organized farmers can benefit from the market by adopting strict quality control measures and obtaining the necessary certification for their goods. The results are in line with Reardon and Barrett (2000) where the authors have indicated that when households acquire expertise in grades and standards, it results in an increase in the market participation by households. However, smallholder farmers have difficulties in meeting market grades and standards in many developing countries leading to exclusion from participating in mainstream agriculture and marketing (De Battisti et al., 2009; Henson and Jaffee, 2008; Henson and Humphrey, 2010). Further, the present study result reveals that when the farmers do not hold certification, the probability of having market linkage is 58 per cent. In some key export markets for certified products at the global level, smallholders are the predominant group of producers (Potts, Van Der Meer, and Daitchman, 2010). However, smallholders are often disadvantaged and rural poverty accounts for about 75 per cent of the world's poverty (Pierre-Marie Bosc et al., 2012). When market conditions are favourable, the High Level Panel of Experts on Food Security and Nutrition of the Committee on World Food Security (Pierre-Marie Bosc et al., 2012) found that smallholders can respond positively. This includes innovation, organization for accessing new market opportunities, upgrading into processing activities and increasing their market power.

In addition, market information is vital to the market participation behaviour of smallholder farmers. Market information allows farmers to take informed marketing decisions that are related to searching for potential buyers, negotiating, supplying necessary goods, monitoring and enforcing contracts. Necessary information includes information on consumer preferences, quality of the produce, price, quantity demanded, market requirements, and opportunities. Further, the source of market information is also very important because it determines the information's accuracy. In the present study, access to information has been set as a dummy variable, where smallholder farmers with access to information through mobile phones takes the value of one and a farmer that has no access to information takes a value of zero. In addition, access to information was expected to influence market participation positively and implies that smallholders with access to information are more likely to participate in marketing as well as make use of formal markets. The results reveal that access to ICT by farmers increased the chances of having market linkage by 3.51 times compared to non-ICT farmers. In addition, when the smallholder does not have access to ICT the probability of having market linkage is only 22 per cent.

Another variable that is closely linked to information availability is access to extension services, such as access to farming advice and knowledge through extension officers who work directly with farmers. Their primary role is to aid the farmers' groups to make better decisions to increase agricultural production. Further, extension officers are constantly armed with the latest techniques and information related to agriculture and they relay this information to farmers and agricultural businesses. In addition, agricultural extension officers often provide consultation with farmers. In these consultations, they give guidance, talks and demonstrations on the latest technologies related to agriculture and on how they can take advantage of such technologies. Further, they also attend seminars and also work with other experts in agriculture to learn more or even develop new methods that could advance production. In the study area among the interviewed respondents, organic farmers were found to frequently visit the extension workers and get more information related to production compared to conventional farmers. The results from the study reveal that with a unit change in the ratio of frequency of extension contact to average frequency of extension contact by the conventional farmer, the probability of that farmer having a market linkage will increase by $[P(1-P)*B]$ is 0.24 or by 24 per cent. Further, for smallholders who do not have an extension contact, the probability of having market linkage is 60 per cent.

Table 3.3   Logistic regression coefficients of the factors influencing market linkage

Variables	B	Sig	EXP (B)
Dairy	2.73	0.009	15.45
Certification	1.88	0.079	6.55
Education	1.35	0.142	3.87
Extension contact	1.03	0.029	2.72
Information communication and technology	3.51	0.001	33.4
Constant	-9.05	0.000	0.00

Notes: Nagelkerke R Square: 0.78, Chi square: 83.86, -2 Log likelihood: 44.34.
Source: Authors' own compilation.

In addition to the organic farmers' efforts to grow organic produce, the state government and NGOs in Karnataka state had also conducted training to enhance the farmers' knowledge on the cultivation as well as on the proper care of organic produce. Such efforts had yielded positive results on the adoption of organic farming in the study area. However, the farmers who had attended the training courses on organic farming, certification, organic fertilizer and manure production had the tendency towards growing and marketing organic produce. Although the primary objective of the training programmes is to develop the farmers' capability in organic farming, the participating farmers are also made aware of the health and financial benefits of raising organic produce. Therefore, by providing the necessary technical know-how and knowledge on health and financial benefits, the training courses and extension contacts play an important and a vital catalytic role in the adoption and marketing of organic produce. Such findings are consistent with the findings of other studies that revealed a significant positive influence of training and extension contact on adoption and marketing of organic produce (Thapa and Rattanasuteerakul, 2011). Overall, for the farmers who have dairy, certification, extension contact and ICT, the probability of having market linkage is 90 per cent.

Conclusions and Recommendations

The study was conducted to identify the factors which influence the market linkage for small holder organic and conventional tomato farmers in Karnataka, India. The research results revealed that access to information, education, extension contact, certification, and owning livestock are the most important factors that influence market linkage for the smallholder farmer. Further, results revealed that conventional smallholder groups need training, facilitation and support services from the government to adopt organic farming, as well as help to connect with markets. There is urgent need to encourage farmers' training, as some small-scale famers prefer sticking to traditional ways of producing and marketing, hence are often excluded from markets. If these farmers have to become a part of the formal agribusiness chain, intervention should start at an individual farmer level. Farmers need to strengthen traditional knowledge with training, particularly, training in management, and technical skills are required by smallholders in order to master the commercial requirements of producing for competitive markets to be able to manage risks and product innovations. Even in the study area where small-scale organic farmers have formed farmer organizations, training is important for modernizing their value chain knowledge and for improving their management and business alliance practices. In the study area, few conventional farmers revealed their lack of knowledge on the importance of certification and value addition, which is the reason for their refraining from such practices. However, organic farmers with a group certification were utilizing value added activities like grading, labelling, and packaging. Therefore, knowledge related to value addition should be disseminated to conventional farmers, because value addition can open up opportunities and increase the farmers' profitability for markets. It is important that the farmers, in cooperation with the private and public sectors, develop and initiate value-adding practices. Practices that do not need a lot of capital like packaging, cutting and drying can be considered by the farmers without outside help. The private and public sector can assist with training farmers about value addition and provide financial assistance especially for the practices that require larger capital commitments. Further, the public and private sectors and their partnership can support the emerging technical innovations for smallholder farmers. They may be in the form of investments in infrastructure such as improved roads, telecommunications and market related infrastructure. Development of such facilities can reduce transaction costs and induce farmers to move towards a commercial agriculture system. In addition, investments in value addition, group certification, technical training, market information systems and collective marketing of tomatoes may provide a potential avenue for enhancing market participation and production of marketed surplus by rural households.

Further, creating sustainable market opportunities to smallholder farmers will provide an incentive for continued production. However, as a way of expanding market opportunities for organic producers in Karnataka, there is a vital need to understand the complexity of the interrelated reasons as to why there has been little growth in the organic production area and market activity in the region.

Future Research

Since the sample in this study is non-representative, that is, only comprised of smallholder farmers growing tomatoes based in south Karnataka, the generalization of the findings should be approached with caution. Given the importance of this issue and the scarce literature available in the country, the same research should be carried out with smallholders growing other types of vegetables from other parts of the country with a wider variety and dimension of samples in order to verify and produce more generalizable results.

References

Aldrich, J.H. and Nelson, F.D., 1984. *Linear Probability, Logit, and Probit Models.* Vol. 45. Sage.

Aldy, J.E. and Stavins, R.N., 2007. *Architectures for Agreement: Addressing Global Climate Change in the Post-Kyoto World.* Cambridge, Cambridge University Press.

Armendáriz, B. and Morduch, J., 2010. *The Economics of Microfinance.* MIT Press.

Bosc, P.-M., Berdegue, J., Goita, M., et al., 2012. Investing in Smallholder Agriculture for Food and Nutrition Security [online]. V0 DRAFT – A zero-draft consultation paper, High Level Panel of Experts on Food Security and Nutrition. Available at: http://www.fao.org/fsnforum/sites/default/files/files/85_Smallholders_v0/HL PE%20V0%20draft%20-%20Investing%20in%20SH%20-%2020-12-2012. pdf. Accessed on 17 September 2012.

Bourn, D. and Prescott, J., 2002. A Comparison of the Nutritional Value, Sensory Qualities, and Food Safety of Organically and Conventionally Produced Foods. *Critical Reviews in Food Science and Nutrition* 42 (1), 1–34.

Casellas, K., Berges, M., and Calá, C.D., 2006. What Determines the Economic Links Among Organic Farmers? Empirical Evidence from Argentina. Poster paper prepared for presentation at the International Association of Agricultural Economists Conference, Gold Coast, Australia, 12–18 August 2006.

Chander, M., Bodapati, S., Mukherjee, R., et al., 2011. Organic Livestock Production: An Emerging Opportunity with New Challenges for Producers in Tropical Countries. *Rev. sci. tech. Off. int. Epiz.* 30 (3), 569–83.

Chouichom, S. and Yamao, M., 2010. Comparing Opinions and Attitudes of Organic and Non-Organic Farmers Towards Organic Rice Farming System in North-Eastern Thailand. *Journal of Organic Systems* 5 (1).

De Battisti, A.B., MacGregor, J., and Graffham, A., 2009. *Standard Bearers: Horticultural Exports and Private Standards in Africa*. IIED.

Duram, L.A., 1999. Factors in Organic Farmers' Decisionmaking: Diversity, Challenge, and Obstacles. *American Journal of Alternative Agriculture* 14, 2–10.

Eyhorn, F., 2007. *Organic Farming for Sustainable Livelihoods in Developing Countries? The Case of Cotton in India*. Vdf Hochschulverlag AG.

Eyhorn, F., Ramakrishnan, M., and Mäder, P., 2007. The Viability of Cotton-Based Organic Farming Systems in India. *International Journal of Agricultural Sustainability* 5 (1), 25–38.

Fairweather, J.R., 1999. Understanding How Farmers Choose between Organic and Conventional Production: Results from New Zealand and Policy Implications. *Agriculture and Human Values* 16 (1), 51–63.

FAO, 2008. Linking Farmers to Market: Some Success Stories from Asia-Pacific Region [online]. Asia-Pacific Association of Agricultural Research Institutions, FAO Regional Office for Asia and the Pacific Bangkok, Thailand. Available at: http://www.apaari.org/wp-content/uploads/2009/05/ss_2008_01. pdf. Accessed on 24 September 2012.

Ferris, S., Robbins, P., Best, R., et al., 2014. Linking Smallholder Farmers to Markets and the Implications for Extension and Advisory Services. MEAS discussion paper series on Good Practices and Best Fit Approaches in Extension and Advisory Service Provision.

Government of India, 2011. Census [online]. Available at: http://www.census2011. co.in/census/state/karnataka.html. Accessed on 16 November 2012.

Guidi, D., 2011. Sustainable Agriculture Enterprise: Framing Strategies to Support Smallholder Inclusive Value Chains for Rural Poverty Alleviation. CID research fellow and graduate student working paper no. 53. Center for International Development at Harvard University, October 2011.

Hellin, J., Lundy, M., and Meijer, M., 2009. Farmer Organization, Collective Action and Market Access in Meso-America. *Food Policy* 34 (1), 16–22.

Henson, S. and Humphrey, J., 2010. Understanding the Complexities of Private Standards in Global Agri-Food Chains as They Impact Developing Countries. *The Journal of Development Studies* 46 (9), 1628–46.

Henson, S. and Jaffee, S., 2008. Understanding Developing Country Strategic Responses to the Enhancement of Food Safety Standards. *The World Economy* 31 (4), 548–68.

Hine, R. and Pretty, J., 2006. Promoting Production and Trading Opportunities for Organic Agricultural Products in East Africa. Capacity Building Study Report 3. United Nations Conference on Trade and Development. UNCTAD, Division on International Trade in Goods and Services, and Commodities; Trade, Environment, Climate Change and Sustainable Development Branch.

Iliopoulou, D., Douma, K., and Giourga, C., 2011. Motives and Barriers to Development of Organic Olive Production. Book of Abstract. International Conference on Organic Agriculture and Agro-Eco Tourism in the Mediterranean.

Jaeck, M., Lifran, R., and Stahn, H., 2012. Emergence of Organic Farming Under Imperfect Competition: Economic Conditions and Incentives. Available at: https://halshs.archives-ouvertes.fr/file/index/docid/793671/filename/WP2012-Nr39.pdf. Accessed on 16 January 2013.

Lobell, D.B., Burke, M.B., Tebaldi, C., et al., 2008. Prioritizing Climate Change Adaptation Needs for Food Security in 2030. *Science* 319 (5863), 607–10.

Markelova, H., Meinzen-Dick, R., Hellin, J., et al., 2009. Collective Action for Smallholder Market Access. *Food Policy* 34 (1), 1–7.

McDowell, J.Z. and Hess, J.J., 2012. Accessing Adaptation: Multiple Stressors on Livelihoods in the Bolivian Highlands Under a Changing Climate. *Global Environmental Change* 22 (2), 342–52.

Morton, J.F., 2007. The Impact of Climate Change on Smallholder and Subsistence Agriculture. *Proceedings of the National Academy of Sciences* 104 (50), 19680–85.

Müller, A., 2009. Benefits of Organic Agriculture as a Climate Change Adaptation and Mitigation Strategy in Developing Countries. Environment for Development discussion paper series.

NHB, 2012. National Horticulture Board Karnataka State Government. Available at: http://nhb.gov.in/.

O'Brien, K., Leichenko, R., Kelkar, U., et al., 2004. Mapping Vulnerability to Multiple Stressors: Climate Change and Globalization in India. *Global Environmental Change* 14 (4), 303–13.

O'Hara, J.K. and Parsons, R., 2012. *Cream of the Crop: The Economic Benefits of Organic Dairy Farms*. Cambridge, MA, Union of Concerned Scientists, UCS Publications.

Odhong, C., Wahome, R.G., Vaarst, M., et al., 2014. Challenges of Conversion to Organic Dairy Production and Prospects of Future Development in Integrated Smallholder Farms in Kenya. *Livestock Research for Rural Development* 26.

Offermann, F. and Nieberg, H., 2000. *Economic Performance of Organic Farms in Europe*. Universität Hohenheim, Institut für Landwirtschaftliche Betriebslehre.

Potts, J., Van Der Meer, J., and Daitchman, J., 2010. The State of Sustainability Initiatives Review 2010: Sustainability and Transparency. International Institute for Sustainable Development.

Puhazhendi, V. and Badatya, K.C., 2002. SHG-Bank Linkage Programme for Rural Poor – An Impact Assessment. Seminar on SHG Bank Linkage Programme at New Delhi, Micro Credit Innovations Department, Nabard, Mumbai.

Quartey, P., Udry, C., Al-hassan, S., et al., 2012. Agricultural Financing and Credit Constraints: The Role of Middlemen in Marketing and Credit Outcomes in Ghana [online]. International Growth Centre, working paper 12/0160. Available at: http://www.theigc.org/project/agricultural-financing-and-credit-constraints-the-role-of-middlemen-in-marketing-and-credit-outcomes-in-ghana-2/. Accessed on 07 January 2013.

Reardon, T. and Barrett, C.B., 2000. Agroindustrialization, Globalization, and International Development: An Overview of Issues, Patterns, and Determinants. *Agricultural Economics* 23 (3), 195–205.

Rezvanfar, A., Eraktan, G., and Olhan, E., 2011. Determine of Factors Associated with the Adoption of Organic Agriculture Among Small Farmers in Iran. *African Journal of Agricultural Research* 6 (13), 2950–56.

Rice, C.W., 2007. Climate Change 2007, Mitigation of Climate Change [online]. Available at: http://www.ipcc-wg3.de/assessment-reports/fourth-assessment-report/.files-ar4/Chapter08.pdf. Accessed on 17 January 2013.

Siegel, P.B. and Alwang, J., 1999. *An Asset-Based Approach to Social Risk Management: A Conceptual Framework*. The World Bank.

Singh, S. and George, R., 2012. Awareness and Beliefs of Farmers in Uttarakhand, India. *Journal of Human Ecology* 37 (2), 139–49.

Thapa, G.B. and Rattanasuteerakul, K., 2011. Adoption and Extent of Organic Vegetable Farming in Mahasarakham Province, Thailand. *Applied Geography* 31 (1), 201–9.

Chapter 4

Breaking the Lock-In to Past Industrial Practices: Triggering Change in a Mature Industry

Diane M. Miller, Frank Calzonetti, Neil Reid, and Jay D. Gatrell

Introduction

The vast literature on industry clusters focuses primarily on innovation regions or sector-specific exemplars with insufficient attention to less successful regions or industries in the mature stage of their life cycle (Scott, 1989; Porter, 1990; Saxenian, 1994; Bathelt and Boggs, 2003). A challenge for such regions is revitalizing mature clusters, perhaps looking for new areas of opportunity that will place it on a new and higher trajectory for future growth and development. Often firms and organizations within lagging clusters are "locked-in" to previously successful routines and practices and are unable to move toward new ways needed to compete at the national or international level. The concept of industrial lock-in is traced to the work of David (1985) and Arthur (1989) and has also been shown to be useful in understanding the behavior of firms in specific regional industry clusters. As noted by Maskell and Malmberg (2007), "proximity matters" and as Cooke (2002) notes, information may be ubiquitous but knowledge is localized. Gertler (2003: 76) makes the distinction between codified and tacit knowledge, and the importance of proximity and "learning-through-interacting" in the transfer of the latter. An industrial cluster *can* go through a life cycle that can both reflect and contribute to the decline of a region but research suggests that there is not a predictable and deterministic straight line path for all clusters (Belussi and Sedita, 2009). It has been shown that external forces or interventions can shock a cluster into reinventing itself so that it can regain its competitive advantage (Elola et al., 2012; Rabellotti, 1999; Meyer-Stamer, 1998). Universities can be a source of intervention and change. Beyond their traditional role of providing a region with educated and informed citizens there are a number of ways that universities can contribute to regional economic development efforts in general and the economic health of industrial clusters in particular. As a neutral stakeholder, universities can function as a hub for innovation, provide an inflow of expertise and new ideas from outside the region, be a convener of regional stakeholders from across municipalities and industries, and contribute to increasing the level of trust and social capital through the process of network weaving (Benneworth and Hospers, 2007; Krebs and Holley, 2006).

This chapter reports on the efforts of a midsized US university and its interventionist role in the Northwest Ohio greenhouse industry cluster. The cluster can be characterized as being path-dependent and locked into physical infrastructure, technology and business practices that are outdated and have been used for generations. The university worked with the growers to first help them organize as a functioning industrial cluster and then introduced leaders within the grower community to an aspirational region to see how new technology, new institutions, and different business models could perhaps revitalize their industry. The example illustrates the difficulty in moving a regional cluster from a "locked-in" position in an old industrial district into a new path that offers the potential of higher economic performance and returns on investment. Even though the participants in the industry and their supporters are able to envision the alternative higher performing position and agreed to work toward a new future, progress is slow and there is reluctance to take more than incremental steps forward. The case of the Northwest Ohio greenhouse industry suggests that long-developed business habits and routines place growers in a comfortable state and moving the industry to a higher level of competitiveness will require additional investment, the likely abandonment of long developed business practices, and the development of new institutions, all of which introduces risk and modifications to established practices. The case also shows how universities can act as external agents to break existing traditions, provide a vision for a transformation of the industry, help transform the industry and create new institutions to stimulate change. The case suggests that "exogenous intervention" may be needed to overcome the inertia of routines, conventions and traditions developed over the course of decades in the region (Martin and Sunley, 2006: 402) and also reinforces the view expressed in "old institutional economics" of the importance of habit and inertia to path dependency and the difficulty of breaking from this past (Hodgson, 1998: 179).

The chapter is organized as follows. We first show how this case fits into the broader work on the topic of breaking the lock-in of clusters in the declining stage of their life cycle. We then provide an overview of the Northwest Ohio industry showing how the university has been intervening to organize the cluster with an aim to revitalizing it. We then provide more detail on the strategy behind identifying Flanders as an aspirational region, the elements of the Flanders greenhouse cluster, and what was learned in visits to Flanders as well as return visits of Flanders personnel to Ohio. The next section reports on technology and practices adopted in Ohio as a result of this intervention. Our last section discusses whether this strategy of engaging an aspirational region can be successful to provide the shock needed to change the path dependency from one of decline toward one of renewed vigor and growth.

Interventions to Break the Lock-In of Late Stage Clusters

The life cycle of a cluster may take it through an establishment stage, a development stage, a maturity stage, and possibly a decline stage (Elola et al.,

2012). While some contributors argue that mechanisms develop that result in clusters losing their advantage over time (Pouder and St. John, 1996), others maintain that the future development of clusters is not subject to historical determinism. Some clusters continually reinvent themselves to avoid decline, generally by innovation driven upgrades and technological advancement and often through their connections to a regional innovation system. The presence of research organizations or universities in support of a cluster can help to stimulate innovativeness and perhaps avoid decline. But even for mature clusters there is evidence that exogenous intervention may be able to break the inertia of routines, conventions and traditions developed over the course of decades (Martin and Sunley, 2006: 402; Hodgson, 1998: 179). A positive shock to the cluster, such as the location of a new multinational corporation bringing its research and innovation networks to a region can renew a cluster (Todtling and Trippl, 2004). A university or other research institution can also serve as a positive change agent.

Institutions and their relationship to convention-based interactive learning in regions have been shown to promote or hinder development of an industrial district (Bathelt and Boggs, 2003; Boschma and Frenken, 2006: 276; Boschma and Frenken, 2011: 302). Once an industrial district emerges institutions can play an important role in their further development and maturation, but they can also limit the exploration of alternative paths forward if they are too tightly linked to older technologies and business practices (Todtling and Trippl, 2004; Belussi et al., 2008). The institutions reinforcing habit and behavior are the business practices, business organizations, and relationship to agricultural land-grant programs that have existed for many decades.

The Northwest Ohio Greenhouse Cluster

The Northwest Ohio greenhouse industry has a rich heritage that can be traced back to the late nineteenth and early twentieth centuries when European (particularly German) immigrants settled in the region. Today, as shown in Figure 4.1, the region (comprising the counties of Erie, Fulton, Lucas, Ottawa, Sandusky, and Wood) is home to 69 greenhouses that have a total of over 4 million square feet under glass or other protection (US Department of Agriculture, 2012). Nearly 50 percent of the region's greenhouses are located in Lucas County, which contains the city of Toledo and is the region's most urbanized county. The average size of greenhouses in Lucas County is considerably larger than those located in surrounding counties. Similarly, the average greenhouse in the region is 82 percent larger than the average greenhouse in the country as a whole (58,965 square feet versus 32,388 square feet) (Table 4.1). The industry's output is dominated by floriculture products such as bedding plants and potted flowering plants that are sold either directly to the consumer or to large retail stores such as Walmart or Home Depot. The industry is dominated by family-owned SMEs that are passed on from generation to generation (Urban Affairs Center, 2009). The majority of the greenhouses in the region today are owned and operated by descendants of those original families. The average age of the region's greenhouse owners is also quite high with the median age being between 50 and 59 years of age. This means that there is a lot of experience vested in the industry's human capital. On the other hand many of these owners are entrenched in old ideas and methods of growing flowers and conducting business. The industry can be characterized as being in the mature stage of the industrial life-cycle and is facing challenges that are impacting its competitive position and calling into question its future viability. These challenges include international competition (particularly from southern Ontario), slow growing markets, ageing infrastructure, outdated technology, and high energy costs (Reid et al., 2009; National Gardening Association, 2013). There is also reluctance on the part of many of the growers to make large investments in new infrastructure and technology. An additional challenge facing the wholesale segment of the industry (those that sell to large retailers) is the seemingly continuous downward pressure on prices being forced upon them by large retail buyers.

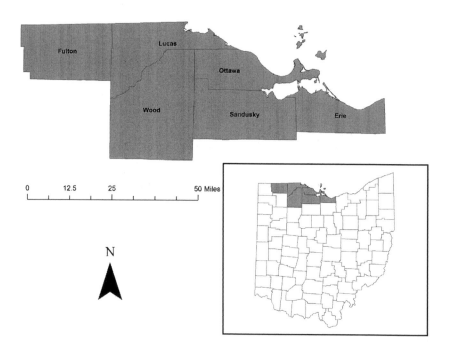

Figure 4.1   State of Ohio and the six counties comprising the Northwest Ohio greenhouse cluster

When university researchers started interacting with the growers and engaging with the industry in the early 2000s they found an industry that also had a significant number of social and organizational impediments to enhancing its competitiveness. The scale of operations was small, growers were focused on local markets, the local supply chain was shallow and poorly developed, there was little inter-grower interaction, a lack of leadership, cohesion, and vision, and a lack of local partners and resources to which the growers could turn to for assistance. In other words, the industry was "institutionally thin" and suffered from "short-termism and localism." Furthermore, it was an industry that was closed to outsiders and impervious to new ideas. A summary of the industry's strengths, weaknesses, opportunities, and threats are provided in Table 4.2. The upshot was that many growers were pessimistic about their future. A 2004 survey revealed that 40 percent of growers felt that the region's industry was going to be less profitable over the next five years while 15 percent indicated that they were planning to either downsize or close their operations during the same time period (Reid and Carroll, 2005).

It should be noted that while the cluster may be considered old by US standards, there are many industrial clusters that are much older, such as the leather cluster of Arzignano, Italy and the ornamental horticulture cluster of Saonara, Italy both of which can be traced to the 1700s or the shipbuilding cluster of the Basque country of Spain that has origins in the twelfth century (Belussi and Sedita, 2009; Elola et al., 2012).

Table 4.1   Number and size of greenhouses, 2012

County	Number of greenhouses	Total square feet under glass	Average square feet under glass
Lucas	34	2,860,928	84,145
Wood	14	778,815	55,630
Fulton	7	376,784	53,826
Erie	6	52,035	8,673
Ottawa	4	Not available	Not available
Sandusky	4	Not available	Not available
Total	*69*	*4,068,562*	*58,965*
United States	26,963	873,290,590	32,388

Note: Total and average square feet under glass are for the 61 greenhouses for which full data are available.
Source: USDA Census of Agriculture, 2012.

Table 4.2   SWOT analysis of Northwest Ohio greenhouse industry

Strengths	Weaknesses
• Critical mass of growers • Extensive grower experience and knowledge • Passionate and committed growers • Large regional production capacity • Access to local university and Agricultural Research Service expertise • Excellent transportation infrastructure and access to markets	• Little history of grower collaboration • Weak brand and market presence • Small size of individual growers • Generational nature of industry • Lack of leadership, cohesion, and vision • Reliance on traditional sources of fuel • Old infrastructure • Dated production technology • Limited access to capital • Thin local supply chain • Lack of global pipelines
Opportunities	**Threats**
• Increase grower collaboration • Develop industry brand and improve marketing • Develop niche markets • Alternative energy options available in region • Water shortages in other regions, particularly, California, create opportunities for Ohio which has good freshwater supplies • Branch out from floriculture • Consumer interest in locally sourced products	• International competition • Big Box store purchasing agreements • High utility costs • Stagnant market

UT Efforts to Trigger Change in the Cluster

The Northwest Ohio greenhouse industry emerged as a target for intervention after receiving the attention of a local and powerful member of the United States House of Representatives. In 2001, the local congresswoman's office identified funding opportunities to support and help grow the industry because of her concern that the industry was in decline and needed a boost that could be provided with the assistance of a federal research agency and local university. She directed funds to the US Department of Agriculture's Agricultural Research Service (USDA ARS) that allowed them locate a small research group at the University of Toledo (UT) to support research and new business practices for the industry. With the support of UT, this research group moved into laboratories on UT's campus and worked with the UT Plant Science Research Center to engage the grower industry and help with notable greenhouse problems of management, design (leading to the creation of the computer program Virtual Grower) and plant health (Frantz et al., 2010). The funding, however, had federal restrictions because it was awarded under the title of Greenhouse Hydroponics. This provided an aspirational direction for the local growers, most of whom were concentrating their efforts in growing ornamentals, as it provided support for them to move toward growing produce. Food production would differentiate the local industry from other regional clusters that focused on ornamentals, relieve some competitive pressure, and help resurrect and rejuvenate the local food industry. Despite the presence of the USDA ARS in Toledo, progress toward local growers adopting best practices and investing in their operations was slow, so the congresswoman directed additional funding to UT through the USDA in 2004 so that university researchers could help organize the growers along the lines of an industrial cluster. This additional funding had both short- and long-term goals. In the short-term the objective was to stabilize the industry and to establish and strengthen its organizational capacity. In the longer-term the hope was to encourage growers to explore, and hopefully adopt, new technology and business practices.

Before the research team could expose the growers to production models that were radically different from what they had been operating for decades they had to enhance the growers' capacity to address some of the more immediate and short-term challenges that they were facing. A team of university researchers, along with a full-time cluster manager hired by the university, met with local growers in October 2004. The goal was to listen to the growers about the economic and business challenges facing their industry and to have a discussion about how the university might be able to provide assistance in meeting those challenges. As the research team listened to the growers they quickly realized that they were dealing with what Lee et al. (2005: 275) refer to as "mobile identities" that comprised both "movers and shakers" and "quiet achievers" (Falk and Kirkpatrick, 2000: 95). Each grower had his or her own unique personality and each had desires and dreams for their particular greenhouse. Collectively, however, the growers wanted to raise the competitiveness of the industry. At the same time, however, they had a strong desire to maintain the family-owned, independent, structure of the industry. Thus many of the conversations that the research team had with the growers became a "process of negotiation between maintaining valued aspects of society, economy, and environment, and engendering new approaches to them" (Lee et al., 2005: 276). It was also apparent that many of the challenges that the growers were facing could best be addressed by them coming together and working collaboratively to identify and implement solutions. However, the low stock of social capital and a low level of trust was a hindrance to such collaboration and meant that collaborative projects had to be carefully chosen and implemented. In cases of limited trust Lorenz (1999: 309) recommends a "step-by-step process" whereby

> [...] firms should start by making small commitments to each other and progressively increase their commitments depending on the quality of the exchange [...] in short, trust was built up through a learning process. Small risks were followed by larger ones, contingent on the success of the cooperation.

The initial collaborative project had to have a number of characteristics. It had to be non-threatening, have a high probability of success, demonstrate the value of collaboration, and enhance the level of local buzz and social capital. The project identified as meeting all these criteria was that of developing a brand identity for the industry and using that brand identity as a platform for subsequent and ongoing collaborative marketing (Reid and Carroll, 2008).[1]

The success of the branding and marketing initiative enabled the research team to be more ambitious in choosing the next collaborative project. A major cost of doing business for the industry is purchasing natural gas to heat greenhouses. Natural gas is purchased on the spot and futures markets and the price paid reflects current and future supply and demand conditions. Growers who made individual purchases of natural gas, spent inordinate amounts of time researching natural gas prices, agonizing over buying decisions, and very often second-guessed themselves once a purchase had been made. A solution to this was to employ the services of an energy consultant who would both bring their market knowledge to the table and then make a single high-volume purchase of natural gas on behalf of participating growers. This would provide economies of scale in purchasing and would relieve the growers of much of the stress associated with individual purchasing. Despite the obvious benefits of this idea, growers were initially hesitant to join the buying group. The hesitancy was based on their unwillingness to relinquish the independence of the buying decision to a larger buying group. However, this hesitancy was overcome thanks in large part to the trust that had been built up during several years of meetings and the positive experience of developing and implementing a branding and marketing campaign.

During the first four to five years of the project the research team's focus was on relationship and trust building within the industry. The collaborative initiatives were quite modest in their goals—stabilizing market share and better managing energy costs—and were not designed to be transformative in terms of industry competitiveness. By mid-2008, however, a number of growers felt that they were ready to take the next step and explore ideas and business practices of a more transformative nature. Since 2004, the university had brought in a number of scholars and experts from across the globe (Australia, Great Britain, California, Wisconsin, Canada, Israel, and Belgium) to visit the university, meet members of the cluster, and also to discuss concepts with the congresswoman or members of her staff. Visiting scholars and experts gave public presentations, visited the research greenhouses at UT and also visited the local greenhouses of industry leaders. These visitors were able to compare and contrast what they saw in Northwest Ohio with the industries that they were already familiar with, giving them a unique authority in the eyes of the local growers. The conversations between the local growers and visitors opened up new possibilities to the local growers, who, in many cases, were still operating the way that their parents had operated. Exposure to experts from other regions was very well received by local growers and functioned as global

---

1   With the assistance of a local branding and marketing firm the growers developed the Maumee Valley Growers brand, a name that reflected the region's location and resonated with local residents.

pipelines that linked growers to new ideas from outside the region (Bathelt et al., 2004). These activities eventually led to interest on the part of cluster leaders to visit other produce and ornamental growing regions, like those in Michigan and Ontario (Canada), and eventually led to discussions on the possibility of organizing some field visits to see the greenhouse industry of Flanders, Belgium. Although there was much progress in transforming a group of growers into a functioning cluster that scored some successes, such as the group natural gas purchasing agreement, the university-ARS team agreed that exposure to an aspirational region may be needed to spur growers into action to invest in new technology and business models. Flanders looked promising as an aspirational cluster.

Comparison of Business Practices and Institutional Support: Ohio to Flanders

The technical leader of the UT based USDA ARS research group provided a connection to a Flemish grower industry contact –a PhD from Ohio State University who worked within the Flanders industry supply chain. This individual provided the ability to develop a long-term partnership with one of the leading areas in the world in greenhouse technology. As UT faculty and ARS scientists looked into the Flanders greenhouse industry, it was decided that investments in UT and ARS personnel trips to the region be undertaken to confirm that Flanders was an appropriate aspirational model for the Northwest Ohio greenhouse cluster. UT sent its own researchers and administrators on two separate trips to study the region and learn more about the food production and distribution system of Flanders and find out if this region's industry was an appropriate analog to provide an aspirational model. These trips reinforced their early hypothesis that this region has one of the most advanced and sophisticated food production and distribution systems anywhere in the world. It was also important to confirm that this was a region not in competition with the Northwest Ohio industry cluster. Table 4.3 summarizes the significant characteristics of Ohio and Belgium that the researchers took into account when initially comparing the two regions.

Table 4.3   A comparison of significant characteristics of Ohio and Belgium

	Ohio	Belgium
Population and land mass	About 11 million in 41,000 square miles	About 11 million in 12,000 square miles
Land	Widely available for cultivation; inexpensive	Less available for cultivation; expensive
Climate	Humid continental, uniform distribution of rainfall, cold winter, hot summer	Temperate maritime; cloudy, uniform rainfall distribution, cool winter, warm summer

Table 4.3   *continued*

	Ohio	Belgium
Groundwater	Highly available, inexpensive (~$0.02/ft³)	More difficult to access, expensive ($0.13/ft³)
Mechanization and labor	Largely manual; hispanic workforce	Automated and manual; Moroccan workforce
Product lines	Primarily floriculture	Primarily vegetables
Production system	Polyculture	Monoculture
Ownership	Family-owned	Family-owned
Irrigation	Well or municipal water sources to runoff	Closed systems utilizing rainwater
IPM strategies and use of biologicals	Adopted by some	Industry standard
Regulations	Low to moderate; industry generally reactive	High; consumer driven; industry proactive

While the salient characteristics of the two clusters are quite different in many respects, including climate, size, and business practices, Flanders offered one of the most important elements for Ohio growers to look to—that is another cluster made up of small family-owned businesses. In the international greenhouse industry, many of the successful regional clusters are made up of corporate farming operations. Since the local Northwest Ohio growers were family run operations (and wanted to stay that way), it was important that any possible aspirational groups also be dominated by family-owned businesses. Thus, the interaction with Flanders resulted in grower to grower conversations on how a family business could adopt the latest technology, produce superior products, and work together to build a strong value chain. It was also critical that the aspirational cluster use of state-of-the-art technology. Since the Northwest Ohio cluster is an aging cluster, much of their current infrastructure and technology is also aging. It was important to introduce growers to a thriving cluster that was making the most of current greenhouse technologies. The aspirational cluster also needed to be supported by a major research center or university, preferably with a research greenhouse. UT's researchers discovered that the Flemish cluster was affiliated with several key research partners, including a local university, a local fertilizer company and a laboratory operated on behalf of the industry that was supported by a combination of funding from the government and from the growers organization.

Beyond these key factors, another component led to UT selecting Flanders as the aspirational region for the Northwest Ohio cluster; the Flemish growers have the largest cooperative vegetable auction in Europe. This auction meant that the group had to be supported by a very sophisticated supply chain. They also had cooperative branding with strong emphasis on product quality and a high level of quality control. This Flemish cluster was also planning for the future. It placed a high value on specialized sustainable agriculture that exceeds current and meets future European Union standards; something that is not yet an issue in the United States, but may be in the near future. This final dynamic of the Flemish cluster made it an ideal aspirational region for the growers from Northwest Ohio to study and learn from.

After identifying the Flemish greenhouse cluster as an appropriate aspirational region, UT identified several leading growers of the Ohio cluster who were interested in the future of the local industry and excited about the opportunity to study an internationally successful model. The first trip, in February 2010, consisted of young growers interested in developing their own greenhouse business and cluster supporters and the researchers who had been working with them. The second trip, in February 2012, took additional growers; these growers were selected because of their ability and desire to make changes to their own greenhouses and provide an example to their industry. On these trips, the organizers stressed the deep supply chain. To emphasize the nature and extent of the supply chain that exists in Flanders, the growers visited both a manufacturer and end-user of fruit grading and packing machinery in 2010. Another aspect that the organizers highlighted was the Flemish growers' monoculture system and specialization. The Flemish growers grew one, or in rare cases, two produce items. Their knowledge of their plants, diseases and pests, watering and care was excellent because their focus was limited. This was completely different from the Northwest Ohio growers, who did not specialize and grow several types of plants including both produce and ornamentals. On both trips the visiting growers were shown the research greenhouses and laboratories associated with the Flemish cluster. These visits gave the growers an opportunity to see how the Flemish growers adopted technology developed and demonstrated at the research station and other laboratories.

The Northwest Ohio growers were interested in other key aspects to the Flanders growing model, some parts of which are potentially transferable while other parts are not. Many of the aspects of the model that are not transferable are different because of the difference in the governments and the availability of government support. The Flemish growers had set up their own branding, much like the Northwest Ohio growers were beginning to do with Maumee Valley Growers. However, the Flemish brand focused on quality and adapting to anticipated stringent environmental and food safety rules and regulations of both the EU and Belgium. The Northwest Ohio growers do not have the same parameters for food safety currently, but do understand that they could in the future. They liked the idea that the growers focus on a monoculture approach: growing either a single crop or related crops (e.g., an entire greenhouse dedicated to cucumbers). In this way, the growers are very specialized. This concept was very foreign to the Ohio growers, growing whatever is in season or that they have a market for in the near future. Another concept that they saw and heard a lot about in Belgium was the growers making huge investments to construct new facilities instead of using or renovating older structures. When the Ohio growers questioned the Flemish growers about why they always build new the Flemish growers explained that there was government funding for new structures, but no government support for using or rehabilitating older structures. The modern greenhouses belonging to Flemish growers allowed for highly technical and mechanized production that minimizes labor costs. This is not an opportunity for growers in the United States, so it is not a model that will work for the Northwest Ohio growers, but it did highlight another key difference between the two industries and pointed to the differences in government support to the industry across the two regions. While Flemish growers benefitted from having modern greenhouse structures designed to maximize growth, temperature control and sunlight, Ohio's growers were still operating for the most part in their grandparents' greenhouses (see figures 4.2 and 4.3).

One other major business consideration for both the Flemish and Ohio growers is energy use and costs, but this concern manifests itself in two very different ways for these two grower groups. The Ohio growers have to be very mindful of their energy use and the cost of the type of energy they are using. On the other hand, the Flemish growers have been encouraged by government subsidies to produce their own energy and capture the carbon created by the process to cultivate their plants. They receive additional funding for the energy that they can sell back to the government for other uses. The Ohio growers visited one greenhouse that produced so much extra energy; they were providing energy to most of the homes within their surrounding community.

Figure 4.2   Example of Northwest Ohio aging infrastructure

Figure 4.3   Example of Flemish high tech/high mechanization infrastructure

The trips also gave the Northwest Ohio growers several opportunities to visit the Flemish growers at their greenhouses. The Flemish large scale, state of the art technology, modern structures were co-located with the family home, something shared in common with the visiting growers. These modern greenhouses also use natural pollination and pest and disease management; something that was completely foreign to the visiting growers from Ohio, but they recognized immediately as something that they could do easily and inexpensively. The Ohio growers also learned about the built in electronic temperature control and water treatment and recycling systems, all monitored by computer and available via mobile phone for the growers to monitor and control remotely. Technology like this is not being used by the Ohio growers because of the age of their infrastructure and the lack of modernization investments that they are willing to make.

Adoption of Flanders Practices and Stimulation of the Ohio Cluster

After the Northwest Ohio growers returned from Flanders, the University research team held a series of debriefing meetings with those who travelled to Flanders and then held informational meetings with those growers in the community interested in the results of the delegations' travel. Many of the delegation commented that the Flemish growers were 20 years ahead of the Ohio cluster. They also were very interested in the Flemish growers' sustainability efforts, use of biocontrols, organic plant media, and the way that the plant media was prepared or "energized"; together these activities showed the Ohio growers what kind of success was possible with a methodical strategy and the knowledge and technology to make it all work together. This awareness combined with their new understanding of how a resilient organized cluster, with strong value chain and government support could not only survive economic downturns but could thrive.

The Flemish group had originated their cooperative and auction in a similarly hostile economic post-World War II environment when economic pressures and manipulative middlemen brokers pitted grower against grower to depreciate the produce market. Organizing, marketing, and cooperating in the auction gave the growers in Flanders the opportunity to take back control of their sales. This was exactly the type of message that the Northwest Ohio growers needed to hear. They had been suffering from competition from one another and had watched several long time industry colleagues go out of business after losing their wholesale contracts with megastores. The market had changed; consumers used to go right to the grower to get plants for the gardens, now increasing volumes are being purchased from the nearest Walmart or Home Depot. Sometimes those plants come from local growers and sometimes they come from another country, depending upon whoever made them available at the lowest price. This pressure was driving small Northwest Ohio ornamental growers out of business. They needed a new plan, a new model and a new product. Growing produce and finding new ways to work collaboratively seemed to make sense, and the growers were seeing how well it had worked for their counterpart growers in Belgium.

Another lesson from the trips to Flanders was the visibility of the greenhouse cluster to government officials and the assistance to the growers from supporting industries that are providing state-of-the-art products and processes. Ohio growers reached out to two Flanders companies for products to be used in Ohio greenhouse operations. One company, DCM had a number of business lines to support agriculture, with a specialization in soil inoculation and media culture that produced organic media of great interest to Ohio growers. The company also maintains an R&D unit, Scientia Terrae, which works with local universities to address problems facing the Flanders greenhouse industry. Another company, Biobest, focused on pollination and biological pest control. Ohio growers were most interested in using bumblebees for pollination. One of the Ohio growers had implemented the use of bumblebees for natural pollination within months of his return from Flanders. His immediate application of a measure that was so easy and inexpensive to emulate also had a profound effect on the other growers in the cluster who started identifying other cost effective measures that they could replicate, such as a different method of plant pruning (e.g., eliminating contiguous tomatoes shoots to keep the main plant stalk strong enough to carry all of the fruit), using low-cost plastic stem stabilizers for maximum yield support, and trying new lower energy cost lighting in their greenhouses. Thus this provides support for Bathelt et al.'s. (2004) observation that information that one cluster firm can acquire through a global pipeline can spill over to other firms in the cluster (Bathelt at al., 2004). While the Ohio growers took advantage of products they saw in operation in Flanders, they could not duplicate in Ohio the close company relationship that existed between growers and suppliers such as DCM or Biobest (or other companies).

One Northwest Ohio grower who traveled to Flanders decided to replicate as closely as possible and on a much smaller scale much of what he had learnt from the trips. This grower was a recent entrant to the industry and had not yet committed to any particular production technology system or product line. As a result of visiting Flanders this grower made the decision to focus on the production of high-quality heirloom tomatoes that would fetch a premium price in the marketplace and would be grown in a monoculture greenhouse environment utilizing much of the growing system techniques used in Flanders.

The visiting growers had also noticed that some of the merchandizing strategies (e.g., denser plants) used in Belgium resulted in product items that were much more attractive to customers. This was another example of how simple changes could help their businesses. The visiting growers had the opportunity to view how the growers in Flanders worked together in order to increase the entire industry's success. On returning, the visiting growers worked with other leaders in their industry to strengthen their existing working groups and find other means of collaborating; this led to the two main grower organizations merging into one.

In order to keep up the association and exchange of ideas, the Northwest Ohio growers invited one of the industry leaders that they had met in Belgium, from the Mechelen Auction (Flanders), to visit Northwest Ohio to offer his expertise and suggestions. Later UT and the growers also had visits from some of the researchers they met in Flanders to discuss future opportunities to collaborate. In general, as a result of the visits to Flanders the growers have more enthusiasm and see a more positive future for their businesses. Since the beginning of this project spearheaded by UT, there has been an increase in interest of the growers' children staying in the business.

Discussion—Breaking Lock-In?

The efforts to revitalize the Northwest Ohio greenhouse industry from being on a trajectory of decline with little new investment or adoption of latest greenhouse technologies or business practices to one with modern structures, deep use of agricultural science, commitment to sustainability, and savvy business practices demonstrates the importance of external shocks to break industry lock-in. The case provides further support that universities and other research organizations (in this case the United States Department of Agriculture) can provide a stimulus to introduce cluster members to new technological approaches and business practices by building trust to allow for the development of a regional system of innovation (Todtling and Trippl, 2004). A number of lessons were learned from this experience. First, reorienting an old industrial district must include efforts to relate new technological orientations and appropriate business methods from the university or research organization *along* with work to build the trust among the cluster's members that is required to use this additional information. In the case of Northwest Ohio, the initial effort was in providing a local source of research but it was not until there was a dedicated effort to bring the growers together into a functioning cross-fertilizing cluster that progress was made. Second, revitalizing a cluster takes time. The effort in Northwest Ohio began in 2001 and even though there is now excitement among the growers and some are investing in new technology in use in Belgium, they still are working in old structures that restrict adoption of some technology. Second, face-to-face interaction with members of an aspiring but not competing cluster showed success. Even though the Flemish cluster was quite different in many respects, the important point was that both the Ohio and Flemish cluster are composed of family businesses who could relate to having a greenhouse operation attached to the family home. The growers from Ohio were able to discuss with Flemish growers issues of family succession and ways to excite the next generation about why they should stay with the business. The Ohio growers and Flemish growers perhaps could relate to one another more easily than a grower would relate to a representative of a large corporate growing operation. Third, the role of a university as a convener is very important in developing communication networks and reaching out to other sources of innovation. In the case of Ohio, the university, working with the USDA, was able to forge new innovation networks and bring a suite of new technologies forward, some of which could be adopted immediately. The role of convener also can result in cluster members taking leadership themselves by articulating a new vision modeled after what they saw was possible.

In summary, breaking the lock-in of an old industrial district is a timely project but is within the scope of the activities of universities. Even though the University of Toledo does not have a college of agriculture, the university could still convene the cluster leaders, connect the cluster to other innovation centers, and assist the growers in creating a new vision that may break a downward trajectory for a regional industry.

References

Arthur, W.B., 1989. Competing Technologies, Increasing Returns, and Lock-In by Historical Events. *The Economic Journal* 99, 116–31.

Bathelt, H. and Boggs, J.S., 2003. Toward a Reconceptualization of Regional Development Paths: Is Leipzig's Media Cluster a Continuation Of or a Rupture With the Past? *Economic Geography* 79, 265–93.

Bathelt, H., Malmberg, A., and Maskell, P., 2004. Clusters and Knowledge: Local Buzz, Global Pipelines and the Process of Knowledge Creation. *Progress in Human Geography* 28 (1), 31–56.

Benneworth, P. and Hospers, G.-J., 2007. Urban Competitiveness in the Knowledge Economy: Universities as New Planning Animateurs. *Progress in Planning* 67, 105–97.

Belussi, F., Sammarra, A., and Sedita, S.R., 2008. Industrial Districts Evolutionary Trajectories: Localized Learning Diversity and External Growth. Paper presented at the 25th Celebration Conference Entrepreneurship and Innovation—Organizations, Institutions, Systems and Regions, Copenhagen, Denmark, June 17–20, 2008.

Belussi, F. and Sedita, S.R., 2009. Life Cycle vs. Multiple Path Dependency in Industrial Districts. *European Planning Studies* 17, 505–28.

Boschma, R.A. and Frenken, K., 2006. Why is Economic Geography Not an Evolutionary Science? Towards and Evolutionary Economic Geography. *Journal of Economic Geography* 6, 273–302.

——— 2011. The Emerging Empirics of Evolutionary Economic Geography. *Journal of Economic Geography* 11, 295–307.

Cooke, P., 2002. *Knowledge Economics: Clusters, Learning and Cooperative Advantage*. London: Routledge.

David, P.A., 1985. Clio and the Economics of QWERTY. *American Economic Review* 75, 332–37.

Elola, A., Valdaliso, J., López, S.M., et al., 2012. Cluster Life Cycles, Path Dependency and Regional Economic Development: Insights from a Meta-Study on Basque Clusters. *European Planning Studies* 20 (2), 257–79.

Falk, I. and Kilpatrick, S., 2000. What is Social Capital? A Study of Interaction in a Rural Community. *Sociologia Ruralis* 40 (1), 87–110.

Frantz, J., Hand, B., Buckingham, L., et al., 2010. Virtual Grower: Software to Calculate Heating Costs of Greenhouse Production in the United States. *HortTechnology* 20 (4), 778–85.

Gertler, M., 2003. Tacit Knowledge and the Economic Geography of Context, Or the Undefinable Tacitness of Being (There). *Journal of Economic Geography* 3, 75–99.

Hodgson, G., 1998. The Approach of Institutional Economics. *Journal of Economic Literature* 36, 166–92.

Krebs, V. and Holley, J., 2006. Building Smart Communities through Network Weaving. Orgnet. Available at: http://www.orgnet.com/BuildingNetworks.pdf. Accessed 16 February 2015.

Lee, J., Arnason, A., Nightingale, A., et al., 2005. Networking: Social Capital and Identities in European Rural Development. *Sociologia Ruralis* 45 (4), 269–83.

Lorenz, E., 1999. Trust, Contract and Economic Cooperation. *Cambridge Journal of Economics* 23, 301–15.

Martin, R. and Sunley, P., 2006. Path Dependence and Regional Economic Evolution. *Journal of Economic Geography* 6 (4), 395–437.

Maskell, P. and Malmberg, A., 2007. Myopia, Knowledge Development and Cluster Evolution. *Journal of Economic Geography* 7 (5), 603–18.

Meyer-Stamer, J., 1998. Path Dependence in Regional Development: Persistence and Change in Three Industrial Clusters in Santa Catarina, Brazil. *World Development* 26 (8), 1495–511.

National Gardening Association, 2013. National Gardening Survey. Williston, Vermont, National Gardening Association, p. 260.

Pouder, R. and St. John, C.H., 1996. Hot Spots and Blind Spots: Geographical Clusters of Firms and Innovation. *Academy of Management Review* 21, 1192–225.

Porter, M., 1990. *The Competitive Advantage of Nations*. New York, Free Press.

Rabellotti, R., 1999. Recovery of a Mexican Cluster: Devaluation Bonanza or Collective Efficiency. *World Development* 29 (9), 1571–85.

Reid, N. and Carroll, M., 2005. Using Cluster-Based Economic Development to Enhance the Economic Competitiveness of Northwest Ohio's Greenhouse Nursery Industry. *Papers of the Applied Geography Conferences* 28, 309–19.

———— 2008. Creating Trust Through Branding: The Case of Northwest Ohio's Greenhouse Cluster. In: Stringer, C. and LeHeron, R. (eds), *Agri-Food Commodity Chains and Globalizing Networks*. Surrey, Ashgate, 175–88.

Reid, N., Smith, B.W., Gatrell, J.D., et al., 2009. Importing Change: Canadian Competition and the US Floriculture Industry. *The Industrial Geographer* 6 (1), 3–19.

Scott, A., 1989. *New Industrial Spaces: Flexible Production Organization and Regional Development in North America and Western Europe*. London, Pion Ltd.

Saxenian, A., 1994. *Regional Advantage: Culture and Competition in Silicon Valley and Route 128*. Cambridge, MA, Harvard University Press.

Todtling, F. and Trippl, M., 2004. Like Phoenix From the Ashes? The Renewal of Clusters in Old Industrial Areas. *Urban Studies* 41, 1175–95.

Urban Affairs Center, 2009. Sustainability Family Style: Documenting the Lives of Growers, Gardeners, and Family Farmers in Northwest Ohio. University of Toledo, Urban Affairs Center. Available at: http://uac.utoledo.edu/Publications/MVGOralHistories-FinalReport.pdf. Accessed 16 February 2015.

US Department of Agriculture, 2012. 2012 Census of Agriculture, Volume 1, Chapter 2: County Level Data, Ohio. Available at: http://www.agcensus.usda.gov/Publications/2012/Full_Report/Volume_1,_Chapter_2_County_Level/Ohio/st39_2_034_034.pdf. Accessed 16 February 2015.

Chapter 5

Rural Development through Strengthened Rural–Urban Linkages: The Case of US Local Food Systems

Becca B.R. Jablonski

Introduction

Among researchers, there is broad consensus that rural development policies in the United States (US) have not met the needs of rural people and communities (Stauber, 2001: 33) and that "policies to improve the disappointing economic performance of rural regions are, by and large, not working" (Porter et al., 2004: 3). Quigley (2002), for example, provides a detailed analysis of economic trends in rural and urban areas from 1970–2000 and finds a drop in total rural personal income, rural per capita incomes at about 70 percent of urban incomes, and increasing disparities in wages and salaries between rural and metropolitan regions. Further, the period between 2010 and 2012 represents the first recorded period of nonmetro population loss in US history, highlighting a growing demographic challenge facing much of rural America. At the community level, this may reduce the demand for jobs, diminish the quality of the workforce, and raise the per capita cost of providing services (Kusmin, 2013: 6).[1]

In search of new opportunities to support US rural communities and economies, many researchers have advocated policies that support relocalized food systems (e.g., Dabson, 2007; Dabson et al., 2012; Feenstra et al., 2003; Gillespie et al., 2007; Jensen, 2010; Kubisch et al., 2008; Marsden et al., 2000; Porter et al., 2004). Accordingly, President Obama and the US Department of Agriculture (USDA) created the Know Your Farmer, Know Your Food (KYF) initiative to "help connect producers with new opportunities in local and regional marketing—and to better inform Americans about the business of agriculture and opportunities to connect with farmers and ranchers." The US Secretary of Agriculture has stated that these initiatives support enhanced rural economic development and farm viability, and thus developing local and regional food systems has become one of the department's priorities (USDA, 2013b). Between 2009 and 2012, the USDA funded over 2,600 projects, mostly through farm bill appropriations (USDA, 2013b).

---

1 Though there are important differences between metro and urban, as well as nonmetro and rural, for the purpose of this chapter they will be treated as interchangeable.

Despite the tremendous growth in local food system activity its distribution has not been uniform (e.g., Jablonski, 2014; Jackson-Smith and Sharp, 2008; Kaufman, 2012; Low and Vogel, 2011). There is strong evidence that successful local food systems involve rural or urban-adjacent farms selling into urban markets; local food systems have become a *regional* development strategy focused on strengthened rural–urban linkages through market interactions. A Kaufman (2012) report shows that farmers' markets are concentrated in metro areas. Lichter and Brown (2011) provide justification for this phenomenon, stating that urban customers are willing to pay higher prices for goods. Aleci (2004) and Aleci and Smith (2011) corroborate this, remarking that the historic central farmers' market in Lancaster, PA has difficulty retaining vendors who are increasingly drawn to better-trafficked farmers' markets in more affluent Philadelphia suburbs. And, Carroll and Jensen (2012), in their case study of Crescent City market vendors, remark that many would not be in business today or as successful if they had not participated in the urban farmers markets.

Studies of rural markets corroborate these findings. Stephenson et al. (2008), for example, cite small market size as being positively associated with market failure. Malone and Whitacre (2012) found that the most rural counties were under-represented in direct-to-consumer sales generally. Schmit and Gomez (2011) and Jablonski et al. (2011) report limited overall vendor sales in their studies of rural markets across northern and central New York, respectively. Even in studies of rural communities demonstrating consumer willingness to pay a premium for locally-grown produce, evidence shows that there are often not enough customers to generate sufficient revenue to overcome costs (Biermacher et al., 2007).

Despite the evidence that there may be enhanced market opportunities for farmers selling in urban markets, there is little research to support the efficacy of strengthened urban–rural linkages as a strategy for rural development. From a global perspective, rural and urban areas are obviously economically linked as there are goods produced in rural areas that are consumed in urban markets. However, in a globalized world, the importance of localized rural–urban linkages becomes much more complex, and is understudied (Dabson, 2007; Holland et al., 2011). Holland et al. (2011), in their work on Oregon, claim that not much is known about the relationship between urban centers and their rural peripheries, and that a clearer understanding of these linkages could support more appropriate policies and strategies for investment. Likewise, Snoxell (2005) states that there is no body of research specifically focused on linkages between communities, nor is there a prevailing analytical framework for understanding these linkages. He calls for more research in identifying linkages between communities and assessing the impact of the linkages on communities. Similarly, Porter (2003) and Porter et al. (2004) call for future research to examine the relationship between rural-urban linkages and rural economic growth.

Using a review of the literature, this chapter examines: What are the rural economic impacts of strengthened rural–urban linkages resulting from localized food systems in the US? The rest of this chapter begins by exploring the rationale for a shift in US rural development policy from *rural* to *regional*, with an emphasis on regional development through strengthened urban–rural linkages and market interactions. Second, this chapter reviews previous research from the US and Canada quantifying the impacts of urban-based initiatives on rural communities. Third, this paper explores how ruralness effects the potential impacts of initiatives based on strengthened rural–urban linkages. Finally, this chapter concludes with areas for future research, and implications for policymakers interested in supporting rural economic development.

Defining "Rural"

Many researchers attribute the challenges with US rural development policy to definitional choices for "rural" (e.g., Brown and Schafft, 2011; Dabson, 2007; Dabson et al., 2012; Isserman, 2005). Brown and Schafft (2011) write that rural has always been defined using a location or "place" approach, which "suffers from a number of inadequacies" (7). First, this approach assumes a clear delineation between urban and rural places. Second, while the "urban" definition is often carefully calculated, the "rural" becomes everything else. Shortall and Warner (2012) concur, "nonmetropolitan/rural areas are [defined as] those that are not metropolitan urban areas" (8).

These rural definitional choices have very important policy impacts. First, they affect how data are collected and viewed. The US Bureau of the Census provides a poignant example. The bureau uses population size and density thresholds to carefully delineate what is urban (metro), and then defines what remains as rural (nonmetro) (Dabson, 2007; Dabson et al., 2012; Shortall and Warner, 2012). Careful analysis of the 2000 census, however, using the metro/nonmetro distinction, results in more than half of all rural residents living in metro counties (Dabson, 2007). The specific needs of these constituents may therefore be masked by the county's metro designation.

Second, the definition of rural establishes the rules regarding which localities can benefit from the USDA's rural development programs.[2] Most rural development programs include maximum population thresholds, above which municipalities are not eligible to participate (USDA, 2013c). As part of the Food, Conservation, and Energy Act of 2008 (the 2008 Farm Bill), the secretary of agriculture was required to report to the Committee on Agriculture of the House of Representatives and the Committee on Agriculture, Nutrition, and Forestry of the Senate on the

---

2   In the US there is no Department of Rural Development. Rural Development is one of 34 agencies and offices housed under the US Department of Agriculture (USDA, 2013a). The fact that agriculture supersedes rural in the layers of federal bureaucracy may be indicative of why the US has never had a systematic approach to rural development policy, and thus why it has not met the needs or rural people or places.

department's various definitions of the term "rural," the definitional effects, and recommendations for improvement. The report acknowledges that within the Office of Rural Development there are:

> arbitrary barriers [...] perpetuating community isolation and less cost-effective economic and community development practices [... For example,] if a regional sewer project encounters a municipality of greater than 10,000 population [...] that community cannot be part of the Rural Development financing application no matter how much sense it might make to project engineers geographically and no matter what the impact of including the larger community might have had on end user rates as fixed costs get spread over a larger number of end uses. (USDA, 2013c: 10)

The definitionally established delineation between urban and rural boundaries also has the effect of supporting policy aimed solely at rural or urban areas, rather than toward broader regions encompassing both rural and urban. This may be important as researchers acknowledge the high degree of connectedness and interdependence between urban and rural America (e.g., Irwin et al., 2010). Lichter and Brown (2011) highlight the "new rural–urban interface" marked by "rapid changes now taking place in rural America and the blurring of rural–urban spatial and social boundaries" (566), as well as the "new scholarly dialogue [...] along the rural–urban divide" (585). Kubisch et al. (2008) underscore the influence that rural and urban places have on each other. And Dabson (2007) writes that "leading thinkers on strategies to achieve greater rural prosperity emphasize the value of strengthening productive ties between rural and urban places" (1).

Though historically there has been few federal programs to create and enhance strategic linkages between rural and urban areas, the USDA (2013c) report provides evidence of changes—particularly in regards to an expanded rural definition with respect to local and regional food systems.[3, 4] In making this change, policymakers and agency staff explicitly acknowledge the "role of more populous areas in providing market opportunities for goods and services provided by rural people" (USDA, 2013c: 10). Both the 2002 and 2008 farm bills expanded access for local and regional food enterprises to loans and loan guarantees under the Business

---

3   There are also other programs that have been recently excluded from traditional rural definitions. Perhaps most notably is the Rural Energy for America Program, which was amended in the 2008 Farm Bill to allow agricultural producers to be eligible irrespective of where their operations are located. For more detailed analysis of the other exceptions, see: USDA, 2013c.

4   The 2008 Farm Bill contains few program provisions that directly support local and regional food systems. However, many existing federal programs benefiting US agricultural producers may also provide support and assistance for local food systems (Johnson et al., 2013).

and Industry (B&I) program, to establish and facilitate the growth of local and regional food markets (as well as to support increased access to "local" food in "food deserts"—i.e., underserved communities). The 2008 Farm Bill reserved at least five percent of B&I funding each year for this purpose, and defined eligible businesses as those selling product within 400 miles of the farm, or within the same State (USDA, 2013c).

This shift in farm bill policy reflects the recommendation of many researchers and practitioners advocating for a move to regionalized policies to support rural development, as well as the desire of many health, consumer, and environmental interest groups to strengthen alternative local and regional food systems (e.g., Dabson, 2007; Dabson et al., 2012; Feenstra et al., 2003; Gillespie et al., 2007; Jensen, 2010; Kubisch et al., 2008; Marsden et al., 2000; Porter et al., 2004). The term "local food systems" has prevailed in the US since the 1990s, marked by increased interest in sustainable agricultural production and alternative food markets (Dimitri et al., 2005; Hinrichs and Charles, 2012). And though, other than with respect to the B&I program, the term does not have a legal definition, it most frequently references geographic proximity between producers and consumers, and, secondarily, social, environmental, or supply chain characteristics (Johnson et al., 2013; Hand and Martinez, 2010; Low et al. 2015; Martinez et al., 2010).

Research on Rural–Urban Linkages

Since the 1990s, there has been a revival of research on rural–urban linkages (Irwin et al., 2010; Tacoli, 1998). Though much more developed in an international context, there is recognition by US researchers (particularly rural scholars) of the complicated interactions and linkages that link rural and urban. As the US agricultural landscape has transformed over the last century, the rural and urban are connected like never before (e.g., Irwin et al., 2010), with traditional boundaries and borders becoming increasingly blurred (Lichter and Brown, 2011). The majority of US focused researchers characterize these linkages as forms of theoretical opportunity, especially for rural areas. Porter et al. (2004) for example note that many economic opportunities available to rural regions involve rural–urban connections.

Exchanges of goods between urban and rural areas are an essential element of rural–urban linkages; one often cited opportunity for adjacent rural–urban regions is strengthened market interactions. The ability of urban consumers to purchase food, feed, fiber, energy and tourism/recreational opportunities from rural areas is a crucial factor in the development of rural areas, reflecting the global trend towards market-led strategies. In this view, government investment in production, distribution, and market infrastructure can be seen as a mechanism to compensate for the market imperfections which are at the root of regional disparities (Dabson, 2007; Davila and Allen, 2007; Tacoli, 1998).

The Empirical Evidence: Dispersion of Economic Impact from Strengthened Rural–Urban Linkages

A literature review revealed six studies that quantify the economic impacts of adjacent rural–urban linkages in the US context. Five of the six studies utilize multiregional input–output (IO) or social accounting matrix (SAM) tools to measure the flows of production and consumption and model the relationships and interdependencies across an economic region (Dabson et al., 2009). By imaging a hypothetical shock (i.e., a policy that increases demand for sectoral output) these tools enable researchers to quantify the strength of inter-industry linkages and assess the impact of the shock in the rural adjacent region. Though IO and SAM are widely used modeling tools, there are not many multiregional assessments as they are more complex than single region models (Hughes and Litz, 1996).

Both Waters et al. (1994) and Searls (2011) measured the impact of rural investment (or disinvestment in the case of Waters et al.) on adjacent urban economies. Waters et al. (1994) studied the impacts from timber harvest restrictions in western Oregon on the Portland, Oregon urban core. Their results predicted that 15 percent of the total regional economic impact of declining timber supply would be felt in the Portland urban core, with most of the impacts coming from reduced household spending rather than from inter-industry sales from the reduced output of wood products. Searls (2011) also reported strong linkages between rural and urban Minnesota, and that investment in rural areas elicited a strong multiplier impact in urban communities. Unfortunately, neither study looked specifically at the impacts of urban investments in the adjacent rural area.

In an effort to better understand the contribution of agriculture to adjacent urban economies, Hughes and Litz (1996) constructed a multiregional IO model of the Monroe, Louisiana functional economic area. Their results show that growth in the urban food processing industry did not imply rapid growth in the adjacent rural areas, despite relatively strong rural–urban linkages. Yet a shock to the agricultural sector elicited a significant impact in the urban economy. Multiplier and impact analysis demonstrated that spillover effects from rural to urban were much larger than the spillover effects from urban to rural. Thus agriculture in rural Monroe, Louisiana can be seen to have a significant impact to the urban core, though the reverse is not true in terms of absolute gains in economic activity to the rural economy.

Dabson et al. (2009) utilized similar methodology to calculate regional economic flows in eight regions of central Appalachia. They found high degrees of connectedness between the urban and rural economies, but that urban economies tended to be more diverse and attract and retain income flows. By contrast, rural economies were more dependent on coal mining (usually with nonlocal ownership), and income flows tended not to stick in the rural economies, but rather were passed on to urban areas or the rest of the world (where the ownership was based). They conclude that regional strategies assuming the effects of investment in urban cores will positively benefit rural areas are unlikely to elicit the desired result in central Appalachia.

Holland et al. (2011) and Lewin et al. (2013) wanted to measure the extent to which Portland, Oregon and its surrounding rural periphery are really interdependent. Utilizing a SAM, the authors found that trade flows over time have decreased. The urban core grew faster than the rural periphery, mainly due to its rapid growth in export goods. The spillover impacts of exports (which they defined as the cross-regional effect of exports from one subregion on sales in the other) generally weakened in both core and periphery, although at a much greater rate in the core. Despite absolute declines in linkage strength since 1982, the authors found that a given change in economic activity in the rural periphery still affects both the core and periphery economies; the urban core is more strongly impacted by a change in the rural periphery than vice versa.

In addition to the IO/SAM assessments, Stabler and Olfert (2009) present an in-depth account of the changing relationship of Saskatchewan's towns and cities. They report that the number of farm families dependent on off-farm income has increased, and that many of the nonfarm jobs are located in cities. Rather than moving for these urban-based jobs, increasingly rural residents choose to commute. Of the 20 percent of Saskatchewan's labor force that commutes to their job, 69 percent are rural dwellers. The relevant piece to note is that while the impact is to slow rural exodus, they find that rural communities continue to decline as they lose businesses. The concentration of high-end services and big box retailers in the major cities attracts shoppers. Rural inhabitants thus largely choose to do their shopping while in the cities for work. Stabler and Olfert conclude that rural and urban communities are connected as never before through rural inhabitants commuting to urban centers for employment, to shop, and to access public services. However, this example demonstrates that though rural–urban linkages have been strengthened, the economic impact is retained in urban areas without trickling out to rural communities.

These studies provide strong empirical evidence that linkages exist between adjacent rural and urban economies. However, without exception, the studies that measured economic flows from urban to rural areas show that they are lower than flows in the reverse direction. Based on this literature review there is no empirical evidence that urban-based economic development initiatives have strong rural economic impacts. Based on these findings, local food system strategies that support strengthened rural–urban linkages may not elicit the strong rural economic impact that policymakers purport.

Local Food and Remote (Lagging) Rural Regions

Up to this point, each of the reviewed studies treats rural–urban (or metro–nonmetro) as dichotomous and rural places as homogenous. However, we know this not to be the case. Drabenstott (2001) writes that one of the most compelling features of the US rural economy is its unevenness. He cites 40 percent of rural counties as capturing nearly all rural economic growth. These "growth havens" share certain key characteristics, including: urban-adjacent (proximity to input and output markets); accumulated human and physical geography; natural endowments (i.e., amenity-rich); and commerce hubs (Drabenstott, 2001; Wu and Gopinath, 2008). On the reverse end of the spectrum, several authors write about so-called "lagging rural regions" in the US and European Union. These regions have particular problems such as geographical remoteness, poor infrastructure, low population density, limited employment opportunities, and poor development capacity (Ilbery et al., 2004). The remainder of this chapter looks at how the literature predicts urban-concentrated local foods markets will impact increasingly bifurcated rural communities.

Within the literature, there is broad agreement that "geographical location has a major impact on rural areas: if a rural area is urban-adjacent then it is likely to be a very different place, with different opportunities compared to a remote rural area" (Shortall and Warner, 2012: 9). There is also consensus that proximity to a metro area is associated with faster growth (e.g., Aldrich and Kusmin, 1997; Ilbery et al., 2004; Hinrichs and Charles, 2012). These findings led Quigley (2002) and Drabenstott (1999, 2001) to predict that whereas some rural areas will become rural hubs for goods and services, others will decline in population and economic activity.

Despite the many claims of ruralness (or remoteness) being associated with slower growth, only one study was identified that quantifies this in the US. Wu and Gopinath (2008) created a theoretical model to analyze the interaction between the location decisions of firms and households and its effect on spatial variations in economic development. Their sample contained 2,635 US counties. Results suggest that remoteness is a primary cause of spatial disparities in economic development, accounting for 76 percent, 85 percent, and 87 percent of the predicted differences in average wage, employment density and land development between the top and bottom 20 percent of counties. They show that counties located in remote areas are less attractive to firms and thus have lower demand for labor. Due to lower labor demands, the wage is lower in remote areas, which leads to lower housing prices and lower demand for land development.

Given the particular challenges of remote rural regions, some researchers in the UK have attempted to analyze the potential for local foods as an economic development strategy. Two studies utilized a Delphi technique to forecast factors likely to influence supply chain development in lagging rural regions in the UK. The Delphi method or technique utilizes a panel of experts to predict (or forecast) something that will happen in the future. Each panel provides their predictions in a first round of discussion, and then is afforded an opportunity to revise their predictions based on the responses of other experts.

In the study by Ilbery et al. (2004), experts concluded that though there are some socio-economic advantages to lagging rural regions from growth in local food system activity, there are important barriers that may impede this type of development. In particular, experts cited the small number and size of "alternative" producers, the restrictive influence of bureaucracy, the shortfall of key intermediaries (i.e., slaughterhouses, transporters, and wholesalers), and a deficit of key physical infrastructure (i.e., roads and railways) as important barriers. Likewise, the study by Henchion and McInytre (2005) found that food supply chain participants closer to the market have more power and influence on supply chain developments. This influence (particularly by retailers) is often reflected in and supported by legislation, and has significant implications for small- and mid-scale producers. For example, many of the experts in their study reported that in retail–producer relationships, the retailer dictated the distributor that the producer was required to use. Many of the experts also noted the benefit of direct sales, but cautioned not to overemphasize them. Despite the reduction in transportation costs and increasing availability of telecommunications infrastructure, it still is not feasible for most producers to ship products, unless they were producing very high value goods.

Conclusion

This chapter considers the extent to which strengthened rural–urban linkages through relocalized food systems have the potential to support rural communities. A thorough review of the literature provides evidence that economic development initiatives focused on strengthened rural–urban through market interactions support urban communities disproportionately to rural ones. Further, rural areas more proximate to urban markets are likely to benefit disproportionately to remote rural regions.

Based on this literature review, there are several key areas for future research. First, though some of the empirical studies looked at specific industries or sectors and how their linkages impact rural or urban economies, most did not. The strength and impacts of each sector's linkages within the economy and in urban and rural regions vary, and thus empirical studies looking specifically at local food system activity are needed. A multiregional IO or SAM model that considers the impact of an increase in final demand for local food in an urban areas on the adjacent rural community may provide additional insight into where and how rural communities can elicit the maximum positive impact from growing interest in local food. Second, are there opportunities for planners to work to ensure that rural communities reap substantial benefit from urban-based local food initiatives? If local food initiatives require market outlets of adequate size, are there specific initiatives that can enable farmers in remote rural regions to more fully participate? Perhaps research will reveal opportunities to support remote rural regions through rural-based aggregation, distribution, and processing facilities initiatives?

As with all economic development initiatives, it is critical to be clear about the intended beneficiary of any policy. To the extent that researchers, policy makers, economic developers and planners are looking for ways to support rural economic development, this chapter provides strong evidence that the physical location of local food initiatives should be careful considered.

Acknowledgments

This project was supported by the Agriculture and Food Research Initiative Competitive Grant No. 2012-67011-19957 from the USDA National Institute of Food and Agriculture. The sponsor played no role in the study design, collection, analysis, interpretation of data, writing of the report, or decision to submit the chapter for publication.

References

Aldrich, L. and Kusmin, L., 1997. Rural Economic Development: What Makes Rural Communities Grow? Washington, DC, US Department of Agriculture, Economic Research Service. Available at: http://ageconsearch.umn.edu/handle/33677 [accessed: February 13, 2015].

Aleci, L., 2004. Central Market and Lancaster's Community Food System: Issues, Challenges, Opportunities. Conference schedule and transcripts. Lancaster Economy Forum: Toward a Research Agenda. Franklin and Marshall College. Available at: http://www.fandm.edu/lec/lec-conferences/lec-conference-2004-2005/lancaster-economy-forum-toward-a-research-agenda/conference-schedule-transcripts/product-networks-and-community-economies/linda-aleci-associate-professor-of-art-and-art-history [accessed: December 13, 2012].

Aleci, L. and Smith, D.T., 2011. Mapping the Eastern Market in Lancaster, PA. A Case Study for Emerging Trends in Farmers Markets and Sustainable Food Systems. *Appetite* 56 (2), 517.

Biermacher, J., Upson, S., Miller, D., et al., 2007. Economic Challenges of Small-Scale Vegetable Production and Retailing in Rural Communities: An Example from Rural Oklahoma. *Journal of Food Distribution Research* 38 (3), 1–13.

Brown, D.L. and Schafft, K.A., 2011. *Rural People and Communities in the 21st Century: Resilience and Transformation*. Cambridge, Polity Press.

Carroll, M.M. and Jensen, J.M., 2012. *Building a Regional Food System: A Case Study of Market Umbrella in the New Orleans Region*. Columbia, MO, RUPRI Rural Futures Lab.

Dabson, B., 2007. *Rural–Urban Interdependence: Why Metropolitan and Rural America Need Each Other*. The Blueprint for American Prosperity Metropolitan Policy Program, The Brookings Institution.

Dabson, B., Jensen, J., Okagaki, A., et al., 2012. *Case Studies of Wealth Creation and Rural–Urban Linkages*. Columbia, MO, RUPRI Rural Futures Lab.

Dabson, B., Johnson, R.G., Miller, K.K., et al., 2009. Rural–Urban Interdependence in Central Appalachia. Discussion Paper: Wealth Creation and Rural–Urban Linkages in Working Regions. New York, NY, Rural Policy Research Institute.

Davila, J. and Allen, A., 2007. Mind the Gap! Bridging the Urban–Rural Divide. *ID21 Insights* 41. Available at: http://discovery.ucl.ac.uk/38/ [accessed: February 13, 2015].

Dimitri, C., Effland, A., and Conklin, N., 2005. The 20th Century Transformation of U.S. Agriculture and Farm Policy. Economic Information Bulletin 3. US Department of Agriculture, Economic Research Service.

Drabenstott, M., 1999. Meeting a New Century of Challenges in Rural America: An Overview of the Challenges Facing Rural America, with Suggestions about the Proper Response of Public Policy. The Federal Reserve Bank of Minneapolis. Available at: http://www.minneapolisfed.org/publications_papers/pub_display. cfm?id=3543 [accessed: February 13, 2015].

———— 2001. New Policies for a New Rural America. *International Regional Science Review* 24 (1), 3–15.

Feenstra, G.W., Lewis, C.C., Hinrichs, C.C., et al., 2003. Entrepreneurial Outcomes and Enterprise Size in US Retail Farmers' Markets. *American Journal of Alternative Agriculture* 18 (1), 46–55.

Gillespie, G.W., Hilchey, D., Hinrichs, C.C., et al., 2007. Farmers' Markets as Keystones in Rebuilding Local and Regional Food Systems. In: Hinrichs, C.C. and Lyson, T.A. (eds), *Remaking the North American Food System: Strategies for Sustainability*. Lincoln, NE, University of Nebraska Press, pp. 65–83.

Hand, M.S. and Martinez, S., 2010. Just What Does Local Mean? *Choices: The Magazine of Food, Farm and Resource Issues* 25 (1).

Henchion, M. and McIntyre, B., 2005. Market Access and Competitiveness Issues for Food SMEs in Europe's Lagging Rural Regions (LRRs). *British Food Journal* 107 (6), 404–22.

Hinrichs, C.C. and Charles, L., 2012. Local Food Systems and Networks in the US and UK. In: Shucksmith, M., Brown, D.L., Shortall, S., et al. (eds), *Rural Transformations and Rural Policies in the US and UK*. Routledge Studies in Development and Society. New York, Routledge, pp. 156–76.

Holland, D., Lewin, P., Sorte, B., et al., 2011. The Declining Economic Interdependence of the Portland Metropolitan Core and Its Periphery. In: Hibbard, M., Seltzer, E., Weber, B., et al. (eds), *Toward One Oregon: Rural–Urban Interdependence and the Evolution of a State*. Corvallis, OR, Oregon State University Press, pp. 79–98.

Hughes, D.W. and Litz, V.N., 1996. Rural–Urban Economic Linkages for Agriculture and Food Processing in Monroe, Louisiana, Functional Economic Area. *Journal of Agricultural and Applied Economics* 28 (2), 337–55.

Ilbery, B., Maye, D., Kneafsey, M., et al., 2004. Forecasting Food Supply Chain Developments in Lagging Rural Regions: Evidence from the UK. *Journal of Rural Studies* 20, 331–44.

Irwin, E.G., Isserman, A.M., Kilkenny, M., et al., 2010. A Century of Research on Rural Development and Regional Issues. *American Journal of Agricultural Economics* 92 (2), 522–53.

Isserman, A.M., 2005. In the National Interest: Defining Rural and Urban Correctly in Research and Public Policy. *International Regional Science Review* 28 (4), 465–99.

Jablonski, B.B.R., 2014. Evaluating the Impact of Farmers' Markets Using a Rural Wealth Creation Approach. In: Pender, J.L., Johnson, T.G., Weber, B., et al. (eds), *Rural Wealth Creation*. New York, Routledge, pp. 218–31.

Jablonski, B.B.R., Perez-Burgos, J., and Gómez, M.I., 2011. Food Value Chain Development in Central New York: CNY Bounty. *Journal of Agriculture, Food Systems, and Community Development* 1 (4), 129–41.

Jackson-Smith, D. and Sharp, J., 2008. Farming in the Urban Shadow: Supporting Agriculture at the Rural–Urban Interface. *Rural Realities* 2 (4).

Jensen, J.M., 2010. *Local and Regional Food Systems for Rural Futures*. Columbia, MO, RUPRI Rural Futures Lab.

Johnson, R., Aussenberg, R.A., and Cowan, T., 2013. The Role of Local Food Systems in U.S. Farm Policy. Congressional Research Service Report 7-5700. Available at: http://www.fas.org/sgp/crs/misc/R42155.pdf [accessed: February 13, 2015].

Kaufman, P., 2012. On the Map: Farmers' Markets Concentrated in Metro Counties. *Amber Waves: The Economics of Food, Farming, Natural Resources and Rural America* 10 (4). Available at: http://www.ers.usda.gov/amber-waves/2012-december/on-the-map-farmers-markets-concentrated-in-metro-counties.aspx#.VN4uccawY8o [accessed: February 13, 2015].

Kubisch, A.C., Topolsky, J., Gray, J., et al., 2008. *Our Shared Fate: Bridging the Rural–Urban Divide Creates New Opportunities for Prosperity and Equity.* Washington, DC, The Aspen Institute. Available at: http://www.aspeninstitute.org/sites/default/files/content/upload/SharedFate-Final_10-20.pdf [accessed: December 21, 2012].

Kusmin, L., 2013. Rural America at a Glance, 2013 Edition. EB-24. US Department of Agriculture, Economic Research Service. Available at: http://www.ers.usda.gov/publications/eb-economic-brief/eb24.aspx#.UyCwuhBV8_e [accessed: February 13, 2015].

Lewin, P., Weber, B., and Holland, D., 2013. Core-Periphery Dynamics in the Portland Oregon Region: 1982 to 2006. *Annuals of Regional Science* 51 (2), 411–33.

Lichter, D.T. and Brown, D.L., 2011. Rural America in an Urban Society: Changing Spatial and Social Boundaries. *Annual Review of Sociology* 37 (1), 565–92.

Low, S.A., Adalja, A., Beaulieu, E., et al., 2015. Trends in U.S. Local and Regional Food Systems. US Department of Agriculture, Economic Research Service. Administrative Publication Number 067. Available at: http://www.ers.usda.gov/publications/ap-administrative-publication/ap-068.aspx [accessed: February 13, 2015].

Low, S.A. and Vogel, S., 2011. Direct and Intermediated Marketing of Local Foods in the United States. ERR-128. Washington, DC, US Department of Agriculture, Economic Research Service. Available at: http://www.ers.usda.gov/publications/err-economic-research-report/err128 [accessed: February 13, 2015].

Malone, T. and Whitacre, B., 2012. How Rural Is Our Local Food Policy. *The Daily Yonder*. Available at: http://www.dailyyonder.com/local-food-policy-it-it-truly-focussed-rural/2012/08/24/4364 [accessed: February 13, 2015].

Marsden, T., Banks, J., and Bristow, G., 2000. Food Supply Chain Approaches: Exploring their Role in Rural Development. *Sociologia Ruralis* 40 (4), 424–8.

Martinez, S., Hand, M., DaPra, M., et al., 2010. Local Food Systems: Concepts, Impacts, and Issues. ERR-97. US Department of Agriculture, Economic Research Service. Available at: http://www.ers.usda.gov/publications/err-economic-research-report/err97.aspx [accessed: February 13, 2015].

Porter, M., 2003. The Economic Performance of Regions. *Regional Studies* 37 (6–7), 545–6.

Porter, M., Miller, K., and Bryden, R., 2004. *Competitiveness in Rural U.S. Regions: Learning and Research Agenda.* Harvard Business School, Institute for Strategy and Competitiveness.

Quigley, J.M., 2002. *Rural Policy and the New Regional Economics: Implications for Rural America.* Available at: http://www.eric.ed.gov/ERICWebPortal/contentdelivery/servlet/ERICServlet?accno=ED473470 [accessed: February 13, 2015].

Schmit, T.M. and Gómez, M.I., 2011. Developing Viable Farmers Markets in Rural Communities: An Investigation of Vendor Performance Using Objective and Subjective Valuations. *Food Policy* 36 (2), 119–27.

Searls, K., 2011. *Pilot Study: Estimating Rural and Urban Minnesota's Interdependencies.* Minnesota Rural Partners, Inc.

Shortall, S. and Warner, M.E., 2012. Rural Transformations: Conceptual and Policy Issues. In: Shucksmith, M., Brown, D.L., Shortall, S., et al. (eds), *Rural Transformations and Rural Policies in the US and UK.* Routledge Studies in Development and Society. New York, Routledge, pp. 3–17.

Snoxell, S., 2005. An Overview of the Literature on Linkages between Communities. Prepared for: Building, Connecting and Sharing Knowledge: A Dialogue on Linkages Between Communities. Infrastructure Canada.

Stabler, J.C. and Olfert, R., 2009. One Hundred Years of Evolution in the Rural Economy. In: Porter, J.M. and Avery, C. (eds), *Perspectives of Saskatchewan.* Winnipeg, University of Manitoba Press, pp. 125–47.

Stauber, K., 2001. Why Invest in Rural America—And How? A Critical Public Policy Question for the 21st Century. *Economic Review* second quarter, 33–63.

Stephenson, G., Lev, L., and Brewer, L., 2008. "I'm Getting Desperate": What We Know About Farmers' Markets That Fail. *Renewable Agriculture and Food Systems* 23 (3), 188–99.

Tacoli, C., 1998. Rural–Urban Interactions: A Guide to the Literature. *Environment and Urbanization* 10 (1), 147–66.

US Department of Agriculture (USDA), 2013a. USDA Agencies and Offices. Available at: http://www.usda.gov/wps/portal/usda/usdahome?navid=AGENCIES_OFFICES_C [accessed: February 13, 2015].

——— 2013b. Fact Sheet: Strengthening New Market Opportunities in Local and Regional Food Systems. Newsroom. Available at: http://www.usda.gov/wps/portal/usda/usdahome?contentid=2013/11/0219.xml&contentidonly=true [accessed: February 13, 2015].

———— 2013c. Report of the Definition of "Rural." Office of the Secretary. Available at: http://agriculture.house.gov/sites/republicans.agriculture.house. gov/files/pdf/reports/USDARuralDefinitionReport.pdf [accessed: February 13, 2015].

Waters, E.C., Holland, D.W., and Weber, B.A., 1994. Interregional Effects of Reduced Timber Harvests: The Impact of the Northern Spotted Owl Listing in Rural and Urban Oregon. *Journal of Agricultural and Resource Economics* 19 (1), 141–60.

Wu, J. and Gopinath, M., 2008. What Causes Spatial Variations in Economic Development in the United States? *American Journal of Agricultural Economics* 90 (2), 392–408.

Chapter 6

A Typification of Short Food Supply Chains and First Insights into Respective Success Factors and Bottlenecks in North Rhine-Westphalia

Luisa Vogt and Marcus Mergenthaler

Introduction

Various food scandals (e.g. BSE, EHEC, and rotten meat) have sensitized the public and helped create a principle of scepticism with regard to conventional food production in recent years (Bánáti, 2011; Van Der Ploeg, 2010). As a consequence, consumer demand is becoming increasingly hybridized with a growing importance of 'green' lifestyles (Pearson, Henryks, and Jones, 2011). Parallel to this, structural change in agriculture has been progressing due to decreasing marginal economic benefits. Against this background, the emergence of an 'integrated territorial paradigm' can be detected as an alternative to the conventional, modern paradigm of the agri-food business (Sonnino and Marsden, 2006). Characteristics of this paradigm are the shortening of the relations between producers and consumers as well as the enhancement of products with added value in the form of information regarding origin and quality in order to regain trust and to re-embed food production in easily understandable chains (Renting and Wiskerke, 2010). The empirical diversity of such short food supply chains (SFSC) is immense and case studies show that there are not any clear demarcations between 'alternative' SFSC and conventional food chains. In any case, SFSCs are assumed to occupy only a small market niche up to now (statistics are not available to prove or disprove this; Kneafsey et al., 2013). Renting and Wiskerke (2010) state that 'existing initiatives remain relatively small and localized and [...] viable dissemination models (either by upscaling or "multiplication") are unclear'. Nevertheless, SFSCs are often perceived as having strong potential for strengthening rural economies (Kögl and Tietze, 2010; Renting et al., 2003).

In the last 15 years, researchers from different disciplines have intensified their efforts to observe, to question, and to understand the development of this new paradigm and of non-conventional food chains: Kögl and Tietze (2010) from the perspective of agricultural economics; Parrott et al. (2002) from the perspective of regional planning; Renting and Wiskerke (2010) from the perspective of rural sociology. However, up to now academic debate is still at the stage of approaching this topic. Whilst there are some empirical studies (Guerrero et al., 2010; Hu et al., 2012), deductive reasoning and terminological considerations form part of the research discourse. They shall shortly be reviewed.

Renting and Wiskerke (2010) reflect on the development of governance mechanisms in the agri-food sector in Europe in the last 60 years and hypothesize a change of governance modes for the present. In the governance mode of the 1990s, food is a commodity which is negotiated on liberalized agricultural commodity markets with civil society forming the demand side of these markets. Since the late 1990s and increasingly in the 2000s, the state – in Europe mainly the EU as agriculture is one of the few policy areas where decision-making has been transferred substantially to the supranational level of the EU – has withdrawn from direct market interventions and price support for several reasons. These reasons ranged from (mainly environmental and ecological) negative side-effects of the once supported agri-food system that were geared for productivity growth, up to growing pressure to limit policy support for agriculture from the negotiations within the framework of the World Trade Organization (WTO). Instead, the European Union with its main instrument, the Common Agricultural Policy (CAP), redefined its role in agri-food governance by launching policies and regulations for improving food safety, food quality and environmental conditions. The term 'multifunctional agriculture' emerged as new paradigm (Coleman et al., 2004). As an illustration, in 1992 the Council Regulation (EEC) No 2081/92 was issued regarding the protection of geographical indications and designations of origin for agricultural products and foodstuffs. For the 2000s, Renting and Wiskerke (2010) see the emergence of a new stage of relations between market, state and civil society. This 'integrative, territorial mode of agri-food governance' is characterized by new actors and new institutions such as public–private partnerships, bottom-up approaches as well as citizens' groups and it crystallizes in SFSCs or alternative food networks (AFNs; Renting et al., 2003). These SFSCs and AFNs have a certain territorial embeddedness in common which is reflected in the aim to capture the added value of specific territorial features or proximity issues.

Furthermore, academic debate pays much attention to sharpen up terms. The term 'integrative, territorial mode' is based upon the French concept of *terroir* in agricultural production (Parrott et al., 2002). The locality or even the specific farm with its surroundings where such food is produced makes up the *terroir* which is considered as a source of quality (Charlebois and Mackay, 2010). The *terroir* does not only include physical-geographic and ecological features but also socio-cultural aspects such as artisanal capabilities and resources. Consumers' knowledge of production spaces or even of the actors involved leads to the denomination of these supply chains as 'short food supply chains'. According to Renting et al. (2003), the adjective 'short' refers to different aspects: on the one hand, SFSCs potentially shorten the long anonymous supply chains of the traditional mode of food production; on the other hand, producer–consumer relations are shortened in the sense of being made more transparent by indicating the origin of food. Besides it is argued that a kind of 'shortening' takes place at the locality of production – the territorial embedding of food production could lead to an increase of environmental responsibility of producers for the area of production. Thus, short supply chains do not necessarily mean the minimization of the number of supply chain actors. Some authors do not share the two latter ideas; instead, they stress the proximity component. Kneafsey et al. (2013: 13) define SFSCs as supply chains where 'foods involved are identified by, and traceable to a farmer', and where 'the number of intermediaries between farmer and consumer should be "minimal" or ideally nil' (i.e. 'direct selling' in marketing terms). The last part of this definition restricts its applicability to only a subset of short or 'alternative' food supply chains as understood by Renting et al. (2003) and others.

Sonnino and Marsden (2006) use the adjective 'alternative' to describe *non*-conventional food supply chains taking the word as commonly understood in everyday language. Renting et al. (2003) criticize the term 'alternative' as unspecific whereas Kneafsey (2010: 179) points out that 'alternative' is used to underline oppositional and normative dimensions such as 'aspirations to "reclaim" ownership of food production, "re-connect" consumers with producers through shorter supply chains, "resist" global capitalism, solve problems of social exclusion and ecological degradation and restore access to healthy food as a human right rather than a commodity'. Mostly, 'alternative' is the descriptor of 'food *networks*'. The preference for this term shows the relational perspective of many agri-food scholars that focusses on processes and also allows integrating non-human actants (Ermann, 2005).

A variation of 'alternative food networks' are 'regional food networks' (Kneafsey, 2010). They emerge when a large part of the links of a supply chain, including retailing and consumption, occurs in a specific region and when involved actors perceive this consciously. Thereby, 'region' is again a relational construct and does not refer to a defined geographical size but to a spatial meso level sharing some similar bio-physical, technological, or legislative characteristics. Kneafsey (2010) cites Lang et al. (2009) who apply the term 'regional' to the whole of Europe. Moreover, in contrast to the *terroir* concept actors in regional food networks do not only trade regional food 'specialities', and thus, the emphasis of this kind of network is not on product features but on process characteristics. A further term is 'regional food' which is used to describe specialty food whose qualities can be attributed to a specific geographic origin (Tregear et al., 2007) and which overlaps with *terroir* products. More often agro-food scholars name this kind of food 'local' – instead of 'regional' – food, at the same time, warning against ideologization of the local scale.

Policy makers, the professional public and even some scholars often associate 'local' with trustworthiness; 'local' is deemed to be preferable. The normative meanings ascribed to 'local food' range from 'ecological sustainability, social justice, democracy, better nutrition, and food security [to] freshness and quality' (Kneafsey, 2010: 179). However, up to now there has been a lack of quantitative evidence demonstrating that the impacts of local food supply chains are solely positive, that they differ from conventional supply chains to a greater extent or that they foster rural development. This holds especially true for the ecological realm (Kneafsey et al., 2013). Sonnino and Marsden (2006) detect a particular need for research in the field of sociocultural impacts. On the one hand, they question the spatial origin of this kind of innovation as well as the location of consumer benefits, and thus they query – in general terms – the social justice as well as, in part, the nutrition dimension. On the other hand, they call for a research focus on power relations in these 'new' food supply chains and on the relationship between 'alternative food networks' and conventional supply chains. It remains questionable, however, if these topics can be addressed for 'alternative food networks' or rather SFSCs as a whole given their heterogeneity.

In their seminal work, Renting et al. (2003) emphasize the large variety of SFSCs and, from this they infer the need for systemization. The authors propose two different dimensions to describe producer–consumer relations of SFSCs in more detail. The first dimension is related to organizational aspects that are linked to time and space. Three categories are distinguished: the first are face-to-face SFSCs marked by direct interactions between producers and consumers (with e-commerce enlarging the spatial component); the second are proximate SFSCs characterized by spatial proximity such as producer fairs and local shops that already require more complex institutional arrangements; and the third category consists of so-called extended SFSCs. Producer–consumer relations are lengthened and can reach global dimensions. In spite of the global scale, Renting et al. (2003) qualify this kind of relation as a SFSC because the territorial embeddedness, the knowledge of the *terroir*, is a significant added value of those products. In order to guarantee the local/regional component, more formal institutional codes such as labels, brands, and formal quality assurance programmes are required. The second dimension concerns the understanding of quality. In the first case, quality is linked to particular features of the place of production or to the production process (ethical considerations could also form an additional benefit). In the second case, quality refers to ecological or natural characteristics. In the real world, Renting et al. (2003) underline this, matters are much more complex with a large number of hybrids and with actors acting on different markets at the same time and, thus, playing different games concurrently. Besides, Parrott et al. (2002) state that the differences in quality definitions can be traced back to 'food cultures'; these can be roughly divided into two categories in Europe – a 'northern' one (the ecological/natural concept) and a 'southern' one (the tradition and artisanal production).

Beyond these differentiation criteria, attempts of categorization are quite rare. For an analysis of their socio-economic characteristics, Kneafsey et al. (2013) regard a sales-based subdivision of SFSCs as helpful. Starting from their definition of SFSCs (the producer–consumer relation should be quite immediate) they distinguish sales in proximity (on farm sales, community supported agriculture, farm direct deliveries, off farm sales to the catering sector, and to the commercial sector) and sales at a distance (e.g. internet sales). It should be added that due to their definition of SFSCs, 'local food' mostly comprises single-ingredient products as farmers are usually not involved in processing. In addition to this classification they suggest differentiating between 'traditional' and 'neo-traditional' SFSCs with the latter using new institutional arrangements. The narrow understanding of SFSCs was appropriate for the scope of Kneafsey's et al. (2013) study by order of the European Commission. In preparation of further CAP reforms, a report on a potential new 'local farming and direct sales labelling scheme to assist producers in marketing their produce locally' (Regulation EU No 1151/92, Article 55) should be compiled. However, this conceptualization has a limited usability for different purposes. On behalf of the Ministry of Agriculture of the Federal State of North Rhine-Westphalia, the authors of this chapter aim to develop starting points for policy support of so-called (literally translated) 'regional marketing initiatives' in North Rhine-Westphalia. Without any broadly accepted definition in the German-speaking area, these initiatives largely correspond to the comprehensive approach of SFSCs according to Renting et al. (2003). The political intention is to move them 'beyond their niches' and therefore, to get to know success factors and bottlenecks of different types of SFSCs; different types in so far as the politically recognized heterogeneity of SFSCs does not allow a joint examination. Thus, we endeavour to contribute to fill the research gap identified by Kneafsey et al. (2013). They recommend strengthening systematic and comparative research in order to detect 'strengths and weaknesses of different types of SFSCs throughout the EU' (Kneafsey et al., 2013: 116).

As a first step, we present an inductive typification developed from a marketing and policy support perspective and based upon the empirical diversity of SFSCs found in North Rhine-Westphalia. In the second step, we survey some key features of the competitive situation for the different types of SFSCs in order to gain more knowledge on success factors and constraints. This latter research is based upon expert interviews with representatives of selected types and a group discussion with further experts to validate these preliminary results.

Typification Criteria

For a deeper understanding of possible starting points for policy support, the variety of SFSCs has to be broken down into categories. The geographical focus of this study is North Rhine-Westphalia, so only SFSCs in North Rhine-Westphalia and here only such initiatives with an actor representing it (e.g. a legal person), are included in the typification process. It is an open question to see how far these cover the different types existing throughout Europe or even in other parts of the world. Based upon deductive reasoning, a type of decision tree has been developed for typification. In the following, this decision tree (see Figure 6.1) will be presented and classification rules will be explained.

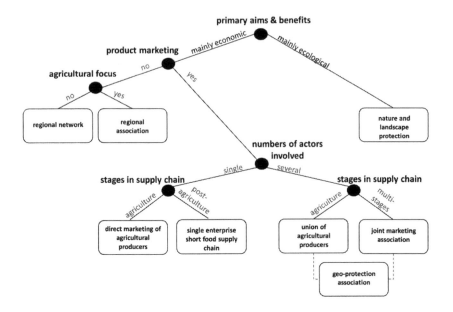

Figure 6.1   Decision tree for typification of SFSCs in North Rhine-Westphalia

Source: Own draft.

Two branches depart from the root, an ecological pathway with a strong focus on nature protection and a socio-economic one. This classification refers to the prevailing aims of actors involved in SFSCs. Emphasis is placed on the descriptor 'prevailing' as actors usually pursue a mix of goals. Initiatives to add value from local 'by-products' of nature and landscape protection usually do not have professionalization of their management and marketing in order to move in large consumer markets as their primary objective. The cultivation of meadow orchards and similar undertakings by local chapters of the Biodiversity Conservation Union Germany and other corporative actors mainly serve nature protection goals. The revenue of selling the processed fruit (e.g. as juice) are normally reinvested in nature and landscape protection activities.

All other SFSCs are labelled as initiatives that aim to generate economic benefits, though to different degrees and via varied pathways. Here, a further differentiation is necessary. A part of these initiatives or rather corporative actors (that have a minimum of organization and formalization such as a common title) are active market players. They are involved in producing, processing and selling local food. The second part of SFSCs actors, however, does not typically deal with product marketing. Rather they work on moderating and networking in order to reduce transaction costs in local/regional food supply chains. Significant transaction costs result from building up new proximate supply chains (Renting et al., 2003).

In many initiatives, public actors play a dominant role as – once established successfully – SFSCs have some features of public goods such as the generation of ecological and social benefits that go beyond the direct economic benefits that accrue to the actors involved. Among the social benefits are the creation of employment possibilities in rural areas and the emerging synergies for other businesses deriving from more regional value adding. Ecological benefits include the preservation of traditional cultural landscapes and possible preservation of natural resources. Thus, especially at the beginning of processes and in situations with indeterminate results, risk-averse private actors are not necessarily interested in investing (time) resources in development processes. However, this constellation does not exclude private actors from being active members of such initiatives.

A sub-division of this branch can occur depending on the industries involved and the importance of agriculture and food to the endeavours to establish local/regional supply chains. 'Regional networks' are typically initiatives whose principal actors engage in regional marketing and focus upon building up local/regional identities. The goal is to foster the competitive situation of local/regional businesses across all industries. Possible additional added value is seen in cross-linking businesses in the specific regions. To that purpose, network activities take place; they are often organized by business development corporations.

The second subdivision is made up by 'regional associations' that have a strong link to the agro-food sector. They are engaged in network activities throughout the agro-food value chain and, partly, undertake tasks in promotion policy. Local Action Groups, responsible bodies at the local level within the framework of the EU support programme LEADER, count among the responsible actors of regional associations. The overriding objective of LEADER is to fulfil the endogenous potential of rural areas; independent of the real economic impact of the agro-food sector and their 'endogeneity', agriculture and related industries are deemed to be the most promising starting points. At the same time, several 'regional associations' in North Rhine-Westphalia comprise private actors as well, or rather are a federation of private actors. Some considerations regarding this issue will be discussed later.

The branch of SFSCs that is actively involved in the market has various ramifications. The first split is done based upon the number of actors who make the strategic decisions regarding the respective SFSCs. Two alternatives can be distinguished: the first applies to SFSCs with a single actor, the second to those with more than one actor. With one single actor, the effort related to interaction with other market players beyond the traditional trade as governance mode is reduced. This branch can further be subdivided based upon the stage of the value chain. The first category refers to primary production; these SFSCs are usually based upon face-to-face producer–consumer relations, i.e. direct marketing of farmers through farmers' shops or selling at local, weekly markets. The other category integrates all kind of individual enterprises involved in SFSCs apart from agriculture. Food retailing companies appear in this group as well as actors at the stage of processing, i.e. food industry companies, provided that they keep local food product lines. SFSCs with several strategically involved actors (and respective higher coordination needs) can again be segmented in two groups; here once more supply chain issues serve as classification rule. The first sub-group, 'unions of agricultural producers', comprises SFSCs where primary producers make the strategic decisions. In the case of 'joint marketing associations', (legal) persons at different stages of the supply chain form the corporative actor and come to strategic decisions.

A variation of these two groups are producer groups, so-called 'geo-protection organizations'. They pursue the goal of registering a product in one of the three EU schemes that promote and protect agricultural products and foodstuffs with a distinctive geographic origin (PDO, protected designation of origin; PGI, protected geographical indication; TSG, traditional specialty guaranteed). These can either be localized at a single stage of the value chain or at different stages. Geo-protection organizations are a special category in so far as common marketing efforts are limited to the development of an application. If the application was successful, they form an alliance to maintain and possibly to defend the protection status.

Issues of quality definition have not been included as a classification rule. However, it can be stated that this group differs from all other types in the explicit understanding of quality as 'spatialized', either by being linked to specific characteristics of the place of production or to locally/regionally specific production processes.

To sum up, eight types of SFSCs could be identified. Not all of them, however, show the potential to create substantial economic multiplier effects or rather to take effects beyond their own operations. Nature and landscape protection oriented SFSCs do not aim for this, from a meso-level perspective the welfare effects of agricultural holdings engaged in direct sale are quite limited, and the effects of 'regional networks' covering different industry sectors on building up SFSCs remain quite vague. Thus, for further analysis of structural characteristics, key features, marketing strategies, and network aspects of SFSCs in North Rhine-Westphalia these three types of SFSCs are excluded.

First Insights Into Spatiotemporal and Organizational Structures, Understandings of Quality, Marketing Strategies, and Network Aspects of Selected SFSCs

Methods

As mentioned in the first section, expert interviews served to gain primary insights into the central features of SFSCs. All actors and SFSCs in this study are located in North Rhine-Westphalia. An initial interview with experts of public administration was used to choose presumed typical and successful representatives of the selected SFSCs types. Then five initiatives (one for each type) were addressed and interviews conducted. The interviews were based on guidelines and contained questions on the context of founding, objectives and organizational issues, on spatial demarcations, natural areas and location issues, on the respective understanding of quality, on marketing strategies, on expectations regarding political support and experiences with it, and finally on interaction issues at the strategic level of SFSCs. These dimensions were selected based upon the literature and deductive reasoning on marketing issues. Interview guidelines were adapted according to the information available about the selected SFSCs via the internet. Four of five interviews were recorded and transcribed. In one case, the interviewee refused permission to record the interview. Thus, we took notes during the interview and completed them immediately afterwards. The analysis of these interviews, summarized in headlines, provided the basis for a group discussion with experts of public administration and further SFSCs. The small group size (four experts, three researchers) allowed for freedom of expression. At the same time, the discussion in a group (and not by single experts) made immediate exchanges possible on contested issues and to raise topics not addressed so far but of interest from the perspective of the experts. In addition, group discussions have the advantage of increasing the validity of results by putting the issues discussed into the relevant context. In the discussion, the preliminary results were validated and confirmed.

In the following section, central features for each type will be presented by way of comparison (see Table 6.1). Each dimension will shortly be explained.

Founding, Objectives, and Organizational Issues

SFSCs emerge from different constellations and motives (Parrott et al., 2002); they may influence market strategies and are thus worth considering. At the same time, the circumstances of their foundation especially are very individual and permit only to a very limited degree of generalization. The joint marketing association sampled was founded bottom-up at the end of the 1990s, originally with the aim of establishing a quality programme for beef. Since then, objectives have been broadened step-by-step and more products have been integrated. Up to now, 55 producers and processors belong to this SFSC. A joint regional label, and thus a more formal institutional code, has been developed in order to create transparency and with this, new distribution channels beyond direct sales. This goal has been reached; meanwhile, the foodstuff is marketed as the local food product line of a food retail company in several supermarkets in the region. Legally, this SFSC initiative is organized as a registered association; membership is restricted to producers and processors. With the help of membership contributions a quality assurance programme is financed. In addition, an advisory board consisting of non-profit organizations and semi-public actors offers expertise and assist in promotion policies and lobbying.

The union of agricultural producers (farmers and gardeners) sampled is also a registered association and was initiated bottom-up at the end of the 1990s.

Table 6.1 Comparison of central characteristics of selected SFSCs

	Joint marketing association	Union of agricultural producers	Geo-protection association	Single enterprise short food supply chain	Regional association
Founding, objectives and organizational issues	Bottom-up pooling of producers and processors in 1998, original objective: establishing a quality programme for beef, since then broadening and integration of more products and introduction of a joint label to permit marketing as the local food product line of a food retail company in the region. Currently 55 members, legal state: registered association. Advisory board with assisting function.	Bottom-up pooling of producers in late 1990s, objective: enlarging customer base of direct sale. Common logo to communicate the direct sale character. Currently 8 market places and 27 members, legal state: registered association.	Top-down pooling initiated by a public actor, objective: quality differentiation via application for legal protection (EU-scheme PGI) of a ham and for this purpose, establishing quality criteria for production. Currently (after protection in 2013) work on quality assurance. Legal state: registered association.	Umbrella brand and product line of a food retail group established in 2011. Labelling as local product, objective: sharing the local food demand.	Top-down pooling initiated by semi-public agencies and regional public actors, objective: strengthening agro-food competitiveness through networking. Currently 66 members, legal state: registered association with a professional agency (currently suffering from financial problems).
Spatial demarcation, natural areas, location issues	Rather informal spatial demarcation (>3 NUTS 3 regions) regarding production and processing, distribution to agglomerations distant up to 100 km. Partly less favoured agricultural area but proximity to agglomerations.	Spatial perimeter for production (and processing): 80 km. Favoured agricultural region and favourite location close to agglomerations.	Sharply defined area of processing (>NUTS 2 region), distribution: no limitation. Favoured agricultural region.	Stretched definition of 'local' (≤NUTS 1 region). Favoured agricultural regions.	Spatial demarcation according to administrative units. Favoured agricultural region.

Understanding of quality	Safety, ecological, and nature-related features (external quality assurance).	Safety, freshness, trust through face-to-face contact (quality control thus only refers to the local origin of food).	Specialty.	Guaranteed provenience.	n/a
Marketing strategies	Product policy: single-ingredient products and processed food. Price policy: middle up to high price segments. Promotion policy: registered brand. No further advertising. Place policy: own distribution centre which allows for new distribution channels (food retail companies).	Product policy: single-ingredient products (partly processed), purchase of additional commodities. Price policy: middle price segment. Promotion policy: communication at the point of sale, advertising in newspapers and on the radio. Place policy: (rather) direct sale on farmers' markets.	No joint marketing.	Product policy: currently only single-ingredient products – critical issue: availability. Price policy: above the price entry-level. Promotion policy: limited to the umbrella brand and one TV spot. Place policies (upstream and downstream): conventional.	n/a
Political support	In an initial phase financial support. Currently no interest in special political support, rather interest in political restraint (rejection of the 'Regionalfenster' label).	Partly financial support for promotion policies. Dependence on favourable municipal decisions about possible locations.	Profit from EU legislation, from subsidies for promotion policy and in the initial phase from knowledge resources of regional public administration.	Participation in a German-wide formal label 'Regionalfenster' initiated by the Federal Ministry of Agriculture. Satisfaction due to opportunity to influence criteria.	In an initial phase financial support for network activities.
Interaction Issues	Win–win situations enable cooperation in spite of vanities and competitive thinking.	Avoidance of too large internal competition through internal clearly defined rules and confidence-building measures.	External moderation in the initial phase to reduce transaction costs. Common economic interests make interaction effective.	n/a	Benefits of networking (without concrete objectives) not enough to create willingness to pay for professional management (possible free rider problem).

Several producers had been involved in direct sale activities beforehand and were interested in enlarging the customer base; the others had been interested in direct sale activities but had not engaged in these due to constructional limits of their farmyards or rather due to their unfavourable location in relation to consumers. Thus, as an alternative to direct sales, a farmers' market has been established, currently with eight market places. A common logo serves as brand and is intended to show that this market is not a retailers' but a farmers' market.

In contrast to both these SFSCs, the geo-protection organization sampled was initiated top-down some years ago. The implementing agency of the Ministry of Agriculture functioned as a moderator to help consolidate organizational structures. The proposal, however, has created interest among businesses of the traditional craft-based food industry as they were aware of the problem of a lack of quality standards. Thus, the unifying interest was to apply for legal protection (the EU scheme PGI) of a meat specialty with the aim of quality differentiation on the market and market leadership in this product segment. The application was successful in 2013. Currently, the geo-protection organization is a self-organized registered association; it works on maintaining the EU label and considers further applications.

The single enterprise SFSC sampled was established in 2011 as an own German-wide umbrella brand and product line of a food wholesale and retail group. Food with local/regional provenience (up to now mainly single-ingredient products such as fruit and vegetables) is labelled as such in order to respond to the perceived growing demand for local food in Germany (DLG, 2011). Completely new supply chains have not been developed; instead existing supply chains are valorized.

The regional association sampled was founded top-down by semi-public agencies and regional authorities; its members (in total 66 'partners') include these bodies as well as several businesses. The association aims to lobby and network throughout the agro-food value chain in a region shaped by intensive horticulture in order to strengthen its competitive situation in a highly competitive market. A by-product of this initiative was the development of a local trademark which is mainly used for labelling cut flowers. This latter kind of activity (which lacks success for different reasons) will not be taken into further consideration as it is rather untypical for a regional association. It has to be mentioned that this regional association (in contrast to all other sampled SFSCs apart from the food retail group) has its own professional agency. At the beginning, its activities were financed by public means but this funding has come to an end. Membership contributions took the place of subsidies. Since then, the agency had to reduce personnel resources.

Spatial Demarcations, Natural Areas, Location Issues

A much-debated issue regarding SFSCs and local/regional food is the geographical size and spatial demarcation of SFSCs. Non-governmental organizations (NGOs) and nature protection agencies tend to give a narrow definition of the spatial layout of SFSCs, including inputs (e.g. feed; Bundesverband der Regionalbewegung, 2013); the main arguments are transparency and credibility issues.

The joint marketing association sampled has fixed in its statutes the spatial demarcation of its supply chain (with the exception of inputs such as seeds and products purchased such as pepper) by a specific landscape yet without defining its borders exactly. The thus rather informal demarcation covers more than three NUTS 3 ('*Nomenclature des unités territoriales statistiques*') regions ('Kreise'). The distributional stages of the supply chain, however, include in addition agglomerations in relative proximity, within up to 100 km. It is argued that consumers in these agglomerations regard the area of origin of these foodstuffs as 'local'.

The area of the union of agricultural producers sampled also comprises more than one NUTS 3 region. Food is produced (at least the main production stages) and processed within a distance of 80 km to the farmers' markets. Customers seem to perceive this distance as sufficiently local as well. As required by the EU, the area of origin is sharply demarked in the PGI case. Based upon documents regarding spatial dimensions of the artisanal tradition of this specialty, the area of the last manufacturing stage has been defined. It covers more than one NUTS 2 area. The market for sales, however, is possibly global. As is usual for geo-protection SFSCs according to Renting et al. (2003), this PGI involves an extended SFSC with an understanding of 'short' not at a spatial distance level.

The single enterprise SFSC sampled has a stretched definition of 'local' food which is oriented towards the criteria of the 'Regionalfenster', a new German-wide voluntary label for local food from the German Ministry of Agriculture which was introduced in 2014. Dependent on customer perceptions and on supply structures, 'local' is defined according to NUTS 1 regions or in relation to large landscapes and it mainly refers to the main ingredient. It is argued that a sufficiently large share of supermarket customers regard this 'localization' as adequate.

The case with the regional association is somewhat different; its spatial demarcation refers to the production context as well as being due to its top-down initiation to administrative borders. The network activities are limited to two NUTS 3 regions. To sum up, dependent on target groups, objectives and market strategies, the SFSCs sampled cover different spatial areas.

A further issue is the natural conditions and the location of SFSCs. Parrott et al. (2002: 243) argue that 'regionally designated products, whilst not exclusive to LFAs [less favoured areas], tend to be associated with agriculturally peripheral regions precisely because such regions have, for a variety of reasons, failed to fully engage with the "productivist" conventions that have predominated in the agro-food system in the second half of the 20th Century'. In North Rhine-Westphalia, LFAs cover almost exclusively the highlands of Sauerland, Siegerland, Eifel, and Teutoburger Wald (MUNLV, 2007).

The SFSCs sampled are, however, mostly situated in the lowlands and so in the most productive areas. This applies to the regional association, the single enterprise SFSC and the geo-protection organization SFSC which is largely located in a favoured agricultural region with intensive livestock production. It also holds true for the union of agricultural producers sampled with its proximate supply chain and its target groups in agglomerations in immediate proximity.

The joint marketing association is the only SFSC which is partly situated in a LFA, a grassland area, whilst it benefits from the neighbourhood of several agglomerations. Thus, at present the simple equation LFA is SFSC does not (possibly does no longer) hold for the situation here. Instead, from a farmer's and processor's perspective the decision to get actively involved in a SFSC and to conduct this kind of quality distinction seems strongly dependent on the location in relation to markets for sales, on operational structures and on the preferences of managers of agricultural holdings and processors.

Understandings of Quality

As several scholars have worked out, quality definitions of local food differ. Products labelled as PGI by the EU scheme such as the specialty of the geo-protection organization sampled only require a regional recipe, i.e. traditional craftsmanship, and a defined area of manufacturing of the final product. The origin of ingredients is thus discretionary. The EU label provides a control system for guaranteeing this kind of quality. Hence, the understanding of quality of this local foodstuff mainly refers to the production process. In contrast to the organoleptic quality dimension of PGI food, safety issues are central to the quality concept of the union of agricultural producers sampled. Freshness (through short supply chains), trust (through direct producer–consumer interaction) but also taste quality is deemed to be the unique selling proposition (USP). The only quality controlled via a (slightly informal) external quality assurance system is the regional origin of food. Further product or process qualities are not an issue.

The additional benefits of foodstuffs produced and processed by the joint marketing association sampled concern safety but moreover ecological and nature-related features. Their cultivation and production guidelines go beyond legal requirements towards organic cultivation; compliance with guidelines is controlled by external supervisors. A further added value is seen in the economic embeddedness; the association relies on the interest of customers in supporting the regional economy.

The single enterprise sampled considers this latter point as important as health, safety, and ecological aspects which are the main associations of customers with local food according to market research. The distinctive quality of the local food product line is in fact limited to the guaranteed provenience. Without any proven positive impacts, the added benefit is thus left to the imagination of customers. As the regional association itself is not involved in product marketing, the question of the understanding of quality does not arise.

Marketing Strategies

Three of the five SFSC actors analysed undertake common marketing efforts. In the case of the geo-protection organization, it is for the members to market the protected product. Thus, member-companies decide on price policies, promotion policies (beyond providing a small promotion set) and place policies (from local to global). The regional association is not involved in any kind of marketing. One of the three SFSC actors with own-product marketing is the single enterprise. Its product range policy comprises the local food product line with a German-wide umbrella brand. As already mentioned, up to now this has included primarily single-ingredient products (vegetables and fruit) but the product range will be expanded to more single-ingredient products (e.g. milk) and more processed foodstuffs (e.g. noodles). Moreover, single outlets of this retail group are encouraged to include the local food of joint marketing associations and others in their product range (without being initiated, controlled, or distributed through the headquarters). In both cases, the most critical issue is sufficient availability. The prices for the local food product line are above entry-level prices up to the middle-price segment with no differences between market locations. There is no special promotion strategy for local foodstuffs beyond the umbrella brand that uses naïve traditional writing and aims to signal trustworthiness. Also at the points of sale the presentation of products does not differ. The most-used advertising method of this enterprise is flyers; they can include food offers. Moreover, thus far one local food TV spot has been produced. As for place policies (upstream and downstream) there is no significant difference compared to conventional supply structures and sale. The structures have not been rearranged; only due to the interest in offering quite consistently local foodstuffs, local intermediaries may feel more secure. The chain thus comprises of the stages of farming, collection, processing, distribution, retail, and ends with consumers.

The chain of the joint marketing association sampled is clearly less extended – dependent on the members it is face-to-face and/or encompasses the stages of farming, possibly processing, distribution centre of local farmers and processors, local food retailers/specialist shops, and consumers. Via their own distribution centre, which at the same time functions as a storage facility and which allows joint sales to food retailers, new channels of distribution have emerged and bargaining power has changed. A single actor now offers a large array of local products; this unburdens food retailers and specialist shops from negotiating with several actors, reduces transaction costs and strengthens the position of local farmers and processors. As mentioned, the local food of this association forms the local food product line of a food retailer in several supermarkets in the region. A possible bottleneck in the case of this association is the potential shortage, especially of processed food, due to the limited number and capacities of the associated craft-based food processors.

The product range includes single-ingredient products as well as a wide range of processed food. Due to multiple additional benefits (see above) products can be positioned in middle to high price segments and target groups demonstrate the respective willingness to pay. The association has a registered brand which is intended to stand for certified quality standards, to reduce communication efforts and to promote recognition. Promotion campaigns beyond the information on product packaging and at events are not conducted mainly due to a rather low advertising budget.

The joint marketing association sampled differs from the union of agricultural producers sampled in three decisive points with one of these related to the supply chain. This union operates a farmers' market (further individual distribution strategies of farmers that range from purchase by wholesalers up to direct sale shall be left out); thus, producer–consumer relations are direct even if, firstly, several farmers purchase part of their products from local producers and processors who are not involved in order to extend the range of products and even if, secondly, farmers employ additional sales staff. Transaction costs are therefore related to different phases of the supply chain; here for example, they mainly refer to negotiations with municipalities for a permit to use favourably located market squares. Dependent on the location, this has already required great effort when the competitive situation with traders' markets and food retailers' markets is tense. The second difference of the analysed farmer-based SFSC compared to the joint marketing association sampled is connected to the understanding of quality and its influence on price and promotion policies. Quality here refers to freshness and safety but not to further ecological benefits, so prices at the farmers' market are at the middle price level (but not above). Promotion policy concentrates on information about the production process in order to emphasize the local provenience of commodities. This takes place directly at the points of sale via screens at market stalls and in addition via open house days. The communication of this USP is seen as a constant challenge as, at a first glance, the difference to traders' markets is not obvious. Further communication instruments with a larger range are adverts in newspapers and via radio. Moreover, the union of agricultural producers relies on word-of-mouth recommendation. The third difference is linked to product policy and distribution policy at the same time. At the market, the various market stalls are potential competitors among themselves. In order to avoid stress, the union carefully chooses the farmers to supply the markets based upon their products, aiming for a large variety. Nonetheless, the product range is too small for a whole market; so, as already mentioned additional commodities are purchased. According to demand, food is more and more processed (potatoes are partly sold pre-cut). No solutions beyond consumer education have been found related to the demand for the whole range of products at all seasons which marks the most obvious difference to conventional weekly markets.

Political Support

Political support refers to three potential fields of intervention: financial support, formal and informal rules at the micro-level, and formal rules at the macro-level. The single enterprise gets political support for its SFSC through the adoption of the 'Regionalfenster' and thus through an official label that signals trustworthiness. The enterprise has influenced the determination of criteria to be applied in awarding this label through membership in the politically initiated association 'Regionalfenster e.V.' and active participation. This lobbying helped to establish quite soft criteria that only focus the origin with a wide understanding of 'local' and that mainly refer to the main ingredient. The enterprise is satisfied with this kind of political support and appreciates the association as an appropriate forum for formulating wishes and expectations. For other SFSCs, the formal rule in the form of the German-wide voluntary label is not as important as for the single enterprise. They rely instead on different institutional codes (such as the geo-protection organization), on less formalized codes or on self-generated rules beyond politics. Nonetheless, the introduction of the 'Regionalfenster' label is partly an issue for some types of SFSCs which regard it (and thus policy action) as a potential threat. Firstly, if this label becomes known among consumers, they might be forced to participate in this scheme as well in spite of the rather high costs for certification and the quality assurance system. The costs are high as smaller SFSCs cannot achieve cost-reducing effects due to comparatively low unit numbers. Furthermore, the joint marketing association sampled and the union of agricultural producers express the fear that consumers might not recognize the difference between the guaranteed origin of the 'Regionalfenster' label and the (usually further reaching) promises of quality of their products. Their USP would thus be lost (cp. Renting and Wiskerke, 2010).

Slightly different formal rules and institutions are vital for the union of agricultural producers sampled. As already mentioned, this SFSC depends on municipal decisions about the use of market squares. Public authorities are moreover partly in demand as funding source e.g. for information and marketing campaigns. According to this SFSC, however, the cost of application (bureaucracy, know-how to fill in application forms, and their own required contribution) make the application for public financial support difficult or even not effective. In the first phase, the regional association sampled got financial support from public means in order to establish a network. In this case the lack of support frameworks for the current stage prevents them from making further applications at the moment.

The geo-protection organization sampled profits from EU legislation; moreover, at the beginning the organization got financial support for communication policy, support in the form of knowledge transfer from public administration for the quite complicated application process at the EU and support in the form of time and knowledge resources in order to initiate the process. In the start-up phase the joint marketing association sampled also got financial assistance in order to successfully establish a quality programme for beef. Currently, this SFSC does not expect any kind of public financial support arguing that relying on their own means results in a slower but at the same time more sustainable and risk minimizing development. Following this argumentation, financial support can not only provoke dead-weight effects but also less conservative marketing strategies.

Interaction Issues at the Strategic Level of the SFSC

At the strategic level of the SFSCs, interaction and cooperation take place within four of the five selected SFSC types. Single enterprises are internally characterized by hierarchical structures; thus, interaction issues are not that relevant. Instead, interaction beyond market and hierarchy occurs within joint marketing associations. The association sampled rates the efficient and effective organization of interaction as crucial; due to the cooperation of independent entrepreneurs, negotiation processes take a rather long time. In the association moreover, vanities and competitive thinking among the members does exist. However, cooperation has worked so far as it has always resulted in win-win situations. This benefit for all participants can be seen as the 'grease' that unifies the association and enables cooperation in the long term.

Representatives of the union of agricultural producers sampled specify the preconditions to keep the network stable: (mainly informal) confidence-building measures have been carried out in order to build up social capital amongst agricultural producers; clearly defined rules help to make decisions; in order to avoid too much internal competition, only producers with a matching range of products are admitted as new members.

The interaction in geo-protection organizations is concentrated on the application for the EU label. As mentioned, in the organization sampled an external moderator from public administration helped to initiate joint action. The same awareness regarding the protection of the specific specialty served as a catalyst. Thus, common interests and the prospect of achieving (economic) benefits provide the basis for effective interaction.

Interaction itself is a core issue of regional associations. In the association sampled, members do not seem to rank the benefits of networking (without any concrete objective) as high enough to pay sufficient membership fees to finance the professional agency with the same staff as at the beginning when public funds subsidized the initiative. A free rider problem may also exist that discourages members from a stronger commitment.

Some Hypotheses on Success Factors and Bottlenecks

Summarizing and based upon some first insights into key features of five SFSC types, the following hypotheses on success factors and bottlenecks for either up-scaling or for multiplication of initiatives can be proposed. Hypotheses on marketing aspects only refer to those SFSCs which market products themselves. Further research is necessary to test them and triangulate the preliminary results.

- Founding: in the case of cooperative SFSCs (i.e. several actors are involved at the strategic level – regional associations, unions of agricultural producers, joint marketing associations), bottom-up initiatives are generally more effective than top-down initiatives.

Top-down founding seems to be successful only in those cases when the potential benefit of cooperation along the supply chain is clear from the beginning and when the impetus of public administration 'only' helps to reduce transaction costs connected with establishing cooperation.

- Spatial and locational issues: in the case of rather small-scale, proximate SFSCs, locational issues in terms of favoured or less favoured areas are not decisive; instead, proximate SFSCs benefit from (perceived) vicinity to markets for sale.

In contrast, spatial and locational issues do not matter at all for single food retailing enterprises with local food product lines showing extended SFSCs and conventional supply structures. Spatial distance to markets for sale is also of rather little importance for geo-protection associations as the companies involved partly deliver to a global market. Moreover, regional associations do not operate exclusively in agriculturally less favourable areas; on the contrary, it may be assumed that efforts to strengthen the competitive situation of agribusinesses through networking mainly take place in favoured areas.

- Product policy 1: local foodstuffs differ in their qualities and USPs. Direct contacts within face-to-face SFSCs may be able to communicate credibly their special qualities; proximate and more extended SFSCs necessitate institutional codes (labels, certification, and quality assurance programmes) and/or rather more formal communication instruments (brands) to convey their added value.

This principle is not new in general; the innovative aspect refers to the experience of two rather proximate SFSCs regarding the challenge of communicating the USP of local food. Proximity alone does not automatically create trust.

- Product policy 2: going beyond specialty foodstuffs (which often creates and occupies their own lucrative market niche), it may be stated that those SFSCs who offer a wide product range and have adequate supplies generate more turnover. Here, economies of scale and economies of scope can be realized.

Unions of agricultural producers may be disadvantaged if they do not buy (or produce) additional and processed products or if they do not move into the specialty food market niche offering foodstuffs such as rare varieties of vegetable, fruit or meat. In addition, the issue of adequate supply is a critical one. On the one hand, the seasonality of agricultural products limits availability. On the other hand, if a specific demand has been created successfully, a lack of satisfaction due to production bottlenecks can frustrate consumers who may choose substitutes.

- Price policy: if local food shows additional benefit in terms of additional qualities and if this benefit is communicated reliably to consumers, local food can secure higher prices. A skimming strategy, however, is successful in the long run only under the premise that the additional benefit is credible.

Thus, (especially extended) SFSCs providing local food which only stands out through a guaranteed origin (without any proof of additional positive effects) have to assert themselves in price competition.

- Place policy: cooperative approaches allow new place strategies as risk is minimized and joint actions such as the bundling of goods reduce transaction costs. Moreover, the following applies: indirect sales foster up-scaling of SFSCs as mass markets can be reached.

If indirect sales are not part of the strategy up-scaling possibilities are quite limited. The move beyond the market niche may only be achieved by multiplication of such SFSCs. A further aspect concerns the transaction costs of indirect sales: a possible bottleneck is the expense of the development of joint bundling and joint marketing compared with single enterprises which use conventional distributional channels upstream and downstream.

- Interaction and cooperation: cooperation along SFSCs does not differ from cooperation in other industries or areas. Cooperation is achieved only if the perceived benefits exceed expected transaction costs. If benefits of cooperation have the characteristics of public goods, private commitment usually does not happen. Moreover, formal rules, explicit contracts, and formal quality assurance programmes communicated through clear labels are indispensable for economically sensitive cooperative actions.

These statements might be rather obvious but in the political context of SFSCs they are not granted. One further aspect might be important: several benefits of cooperation increase parallel to the number of involved actors. On the other hand, the more actors participate, the higher the transaction costs. Above a certain size, professional management is indispensable which, however, again augments cooperation costs. By comparison, single actors profit from considerably lower coordination costs.

Conclusion

The development of a typification scheme for SFSCs has been an attempt to structure and categorize a high degree of variability in the organization of SFSCs. In a limited geographic scope within the state of North Rhine-Westphalia in Germany the proposed typification has proven to be applicable and thus allows a more structured analysis of SFSCs. The first insights yield some interesting results regarding commonalities and differences in the SFSC-types identified. With regard to founding and organizational issues of SFSCs, differentiation between bottom-up and top-down approaches is an important feature but also size with regard to the number of actors involved. The special demarcation of SFSCs can be either more informal or strictly oriented towards very formal, public administrative units. Understanding of quality varies by the inclusion of ecological and nature-related aspects in addition to the mere location aspects. Also differences in the marketing strategies have become evident: regarding product policy, there are differences in the product portfolio with regard to single-ingredient vs. processed products. Also pricing and communication strategies differ. Finally there is a difference in the amount and the timing of political support measures.

With regard to the hypotheses and bottlenecks that have been derived it becomes clear the development of concrete policy advice remains challenging. The bottom-up initiatives show that political support programmes for the establishment of SFSCs are not a necessary condition for success. Rather it can be stated that simply introducing policies that would hamper the development of such initiatives should be avoided. These could be policies that promote strong market integration and market concentration, i.e. policies should be supported that give initiatives the freedom to organize themselves and support some type of formal registration (e. g. administration to register as a legally registered association should not be bureaucratic).

Whenever political actors take initiatives for the foundation of a regional marketing initiative, clear and transparent criteria are necessary in order not to undermine self-organizational forces from within a sector. One criterion could be strong competition among economic actors that hinders the realization of common interests in regional marketing ('prisoner's dilemma'). These could be the case when there is a lack of communication platforms like functioning industry association or when potential partners for a SFSC have very divergent company sizes.

Whenever single companies (but to a lesser degree also strong initiatives as a whole) initiate and control SFSCs, they have incentives to maximize the company's (or companies') profits at the expense of societal expectations. If consumers are not in the position to exert their influence through their buying decision due to information asymmetries ('lack of market transparency'), policies are required to re-establish market transparency so that consumers can make informed buying decisions. More specifically, if the criteria of what a single company defines as 'regional' does not meet consumers' expectations and if there is potential consumer fraud, policies are required to define which products can be labelled as 'regional' and for which products this should not be possible. If consumers are not in a position to make a judicial case (due to dispersed interests), support for consumer protection organization might be an option.

Our results so far clearly show the statement already made by Renting and Wiskerke (2010) that 'existing initiatives remain relatively small and localized [...] and viable dissemination models (either by upscaling or "multiplication") are unclear' still holds true. It lies in the nature of SFSCs that they are highly time- and context-specific. Thus, an open question to be further researched refers to the applicability of the typification beyond the scope of the study presented here. Can the types of SFSCs identified here also be found in other places? Are there any types of SFSCs that cannot be integrated into the proposed typification scheme? For this purpose deriving clear and measurable distinction parameters that could be easily applied to describe a high number of SFSCs would be advisable. With this kind of data and with a relevant sample size more formal and statistical typification approaches could also be employed (e.g. cluster analysis).

Moreover, extending the geographic focus of analysing existing SFSCs could generate additional success factors that were not possible to identify so far. This could be expected in contexts with completely different framework conditions for which variation in the current context has been too small. Analysis of SFCS in different cultural, climatic, geographic, socio-economic and political contexts could be conducted. Increasing the variation in the sample would allow further success factors to be derived and thus policy advice that goes beyond of what has been derived so far.

In addition, if and to what extent the hypothesized success factors are time and location specific or if they can be generalized, has to be tested. For this purpose, more quantitative research approaches with a wide geographic scope that can validate the hypotheses generated here might be advised. This should include the development and validation of operational measuring tools that could be easily implemented with a large number of different SFSC-initiatives.

Acknowledgements

The authors wish to acknowledge the financial support of the Ministry for Climate Protection, Environment, Agriculture, Nature Conservation and Consumer Protection of the German State of North Rhine-Westphalia for the study 'Erfolgsfaktoren und Schwachstellen der Vermarktung regionaler Erzeugnisse' (grant number 17-02.04.01-5/2013). They would also like to thank Janina Wiesmann and Wolf Lorleberg for collaboration, discussion, and constructive comments within the project.

References

Bánáti, D., 2011. Consumer Response to Food Scandals and Scares. *Trends in Food Science and Technology* 22 (2), 56–60.

Bundesverband der Regionalbewegung, 2013. Interner Kommunikationsprozess zur Entwicklung der Auszeichnungs- und Prüfkriterien 'Regional mit Qualität – umweltverträglich, wirtschaftlich'. Unpublished document.

Charlebois, S. and Mackay, G., 2010. Marketing Culture through Locally-Grown Products: The Case of the Fransaskoisie Terroir Products. *Problems and Perspectives in Management* 8 (4).

Coleman, W.D., Grant, W.P., and Josling, T.E., 2004. *Agriculture in the New Global Economy*. Edward Elgar Publishing.

DLG, 2011. Regionalität aus Verbrauchersicht. Deutsche Landwirtschaftsgesellschaft, Frankfurt.

Ermann, U., 2005. Regionalprodukte – Vernetzungen und Grenzziehungen bei der Regionalisierung von Nahrungsmitteln (Sozialgeographische Bibliothek 3). Stuttgart, Franz Steiner Verlag.

Guerrero, L., Claret, A., Verbeke, W., et al., 2010. Perception of Traditional Food Products in Six European Regions Using Free Word Association. *Food Quality and Preference* 21 (2), 225–33.

Hu, W., Batte, M.T., Woods, T., et al., 2012. Consumer Preferences for Local Production and Other Value-Added Label Claims for a Processed Food Product. *European Review of Agricultural Economics* 39 (3), 489–510.

Kneafsey, M., 2010. The Region in Food – Important or Irrelevant? *Cambridge Journal of Regions, Economy and Society* 3 (2), 177–90.

Kneafsey, M., Venn, L., Schmutz, U., et al., 2013. Short Food Supply Chains and Local Food Systems in the EU. A State of Play of their Socio-Economic Characteristics. JRC Scientific and Policy Reports. Luxembourg, Publications Office of the European Union.

Kögl, H. and Tietze, J., 2010. *Regionale Erzeugung, Verarbeitung und Vermarktung von Lebensmitteln*. Universität Rostock, Rostock.

Lang, T., Barling, D., and Caraher, M., 2009. *Food Policy: Integrating Health, Environment and Society*. Oxford, Oxford University Press.

MUNLV (Ministerium für Umwelt, Naturschatz, Landwirtschaft und Verbraucherschutz des Landes Nordrhein-Westfalen), 2007. NRW-Programm Ländlicher Raum 2007–2013. Düsseldorf, Ministerium für Umwelt, Naturschatz, Landwirtschaft und Verbraucherschutz des Landes Nordrhein-Westfalen.

Parrott, N., Wilson, N., and Murdoch, J., 2002. Spatializing Quality: Regional Protection and the Alternative Geography of Food. *European Urban and Regional Studies* 9 (3), 241–61.

Pearson, D., Henryks, J., and Jones, H., 2011. Organic Food: What We Know (and Do Not Know) About Consumers. *Renewable Agriculture and Food Systems* 26 (2), 171.

Renting, H., Marsden, T.K., and Banks, J., 2003. Understanding Alternative Food Networks: Exploring the Role of Short Food Supply Chains in Rural Development. *Environment and Planning A* 35 (3), 393–412.

Renting, H. and Wiskerke, H., 2010. New Emerging Roles for Public Institutions and Civil Society in the Promotion of Sustainable Local Agro-Food Systems. In: Darnhofer, I. and Grötzer, M. (eds), *Building Sustainable Rural Futures. The Added Value of Systems Approaches in Times of Change and Uncertainty*. Vienna, University of Natural Resources and Applied Life Sciences, pp. 1902–12.

Sonnino, R. and Marsden, T., 2006. Beyond the Divide: Rethinking Relationships between Alternative and Conventional Food Networks in Europe. *Journal of Economic Geography* 6 (2), 181–99.

Tregear, A., Arfini, F., Belletti, G., et al., 2007. Regional Foods and Regional Development: The Role of Product Qualification. *Journal of Rural Studies* 23 (1), 12–22.

Van Der Ploeg, J.D., 2010. The Food Crisis, Industrialized Farming and the Imperial Regime. *Journal of Agrarian Change* 10 (1), 98–106.

PART III
Alternative Functions for Rural Areas

Chapter 7

Renewable Energies and Rural Development in Germany: Business Finance and the Role of Trust Illustrated by Two Case Studies from Brandenburg

Sabine Panzer-Krause

Introduction

In order to realize the energy transition in Germany, the German government has focused on the expansion of renewable energies, especially solar energy, wind energy, and energy from biogas. The target by year 2050 is for 60 per cent of all energy demand in Germany to be provided by renewable energies (BMU, 2011). But it is not only since the nuclear catastrophe in Fukushima in 2011 that renewable energies are at the centre of this energy transition process. In fact, since the 1990s a systematic development of renewable energies has been pursued, for example by promoting the development of new energy technologies or by providing a protected market niche by means of enacting the Renewable Energies Law (Erneuerbare-Energien-Gesetz), which subsidizes the power input from renewable energies into the national electricity network. Accordingly, the relevance of the renewable energy industry has risen enormously. Germany has acquired the role of a pioneer in the international context, and the German government calls the green economy, and especially the renewable energy sector, the engine of sustainable development (e.g. BMU and BDI, 2013).

However, this transition process has also led to problems concerning the acceptance of renewable energy production sites (Klagge, 2013). While most of the energy demand arises in urban areas, renewable energy production sites have primarily been erected in rural regions. A number of studies showed that public participation and benefits through job creation help to raise the levels of acceptance (e.g. Pohl, 2013; Wotha, 2013). Concerning job creation, the access to finance for renewable energies-focused businesses in rural regions is an important prerequisite. However, many banks, as the most important suppliers of finance to small and medium sized enterprises (SME) in Germany, have withdrawn more and more of their branches from rural regions during the last two decades. In this chapter, I deal with the relationships between rural SMEs working in the renewable energies sector and their house banks, and I discuss possible modes of interaction and the role of trust from a spatial perspective. By means of two case studies from sparsely populated regions in Brandenburg, I investigate whether such companies in rural areas are disadvantaged concerning their supply of finance and might even be at risk of financial exclusion. First, I will examine the German banking system and structural changes within the banking sector as well as their relevance for the financing of renewable energies-focused SMEs in rural regions. Following, I will outline the theoretical framework and the methodology of the study. Then I will present the two case studies before I draw a final conclusion.

Financing Renewable Energies-Focused SMEs in Germany

Capital and Financing from a Spatial Perspective

As there is a growing consensus that financial markets and their actors have a massive influence on the development of economies and societies (e.g. Musil and Zademach, 2012; Marshall, Pollard, and Wray, 2010; Clark and Wójcik, 2007; Sassen, 1991), Pike and Pollard (2010: 29) argue for embedding finance 'into the heart of economic geographic analysis'. However, French, Leyshon, and Wainwright (2011: 798) point to the fact that research on the concept of financialization has still 'been insufficiently attentive to space and place'. In this chapter, I intend to emphasize the role of financial aspects in supply chains and their spatial dimension in order to make a contribution to the broad body of research on the governance of value chains (e.g. Gereffi, Humpfrey, and Sturgeon, 2005; Bair, Gibbon, and Ponte, 2008). Yet, in this chapter, I limit my analysis to sole actors within supply chains in the renewable energy sector and the relationships to their house bank.

In Germany, SME finance is mostly provided by bank credits. However, since the 1980s the international financial markets have been in a process of restructuring. This process has mainly been driven firstly, by new information and communication technologies, secondly, by market liberalization leading to an internationalization of financial players and intense international competition, and thirdly, by the transformation from a sellers' to a buyers' market entailing the rising demands of clients (Klagge, 2010; Panzer, 2008; Pieper, 2005). In Germany – as in other countries – this has resulted in a fundamental reorganization of the national banking sector.

On the one hand, there have been numerous mergers and acquisitions and thus an institutional concentration. Figure 7.1 illustrates the development of the number of credit institutions in Germany between 1990 and 2012 while differentiating between private banks, savings banks, and credit unions. It shows that there has been an enormous overall reduction of the number of banks. While in 1990 4,527 banks were registered in Germany, this number sank to 1,928 credit institutions in 2012. This corresponds to a reduction of 57.4 per cent. The biggest reduction can be identified in the credit unions sector, where many mergers have taken place. But also the number of savings banks sank significantly from 781 in 1990 to 432 in

2012, whereas the number of private banks has been relatively stable throughout this period with even a slight rise of the number of banks during the last few years.

On the other hand, a strong geographical consolidation of branch networks has taken place in Germany, and many bank branches have been closed (Figure 7.1). Since 2005 alone, when branches of Deutsche Post AG were included in the statistics, 31.4 per cent of all bank branches have shut down. Regional statistics, however, are only available until the year 2003, since then there has been no obligation for credit institutions to report the locations of their bank branches, anymore. Nevertheless, the regional development of the bank branch density (referring to the area per bank branch) in the German federal states between 2000 and 2003 already shows a decline in all parts of the country (Deutsche Bundesbank, 2004). In 2003, the lowest bank branch densities were found in Brandenburg and Mecklenburg-Western Pomerania (Figure 7.2). This situation has worsened since. In the international context, German bank branch density can be compared with bank branch densities in Belgium, the USA or Great Britain, whereas countries such as Italy, Switzerland, France, or Austria are characterized by much higher bank branch densities (Deutsche Bundesbank, 2012).

The structural changes within the financial system have also changed the relationships between businesses and their banks. Traditionally, the credit business in Germany was characterized by the so-called relationship banking which refers to a mode of interaction that is characterized by a strong bond between businesses and their house banks. The purpose is that information asymmetries and thus uncertainty between the partners can be reduced more easily as credit institutions have easy access to business information, especially soft information.

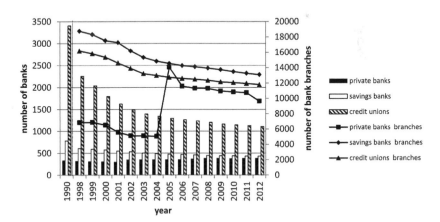

Figure 7.1   Development of the number of banks (1990, 1998–2012) and the number of bank branches (1998–2012) in Germany

Note: Since 2005 Deutsche Post AG branches have been included in the statistics.
Source: Deutsche Bundesbank, 2000, 2004, 2008, 2012, 2013.

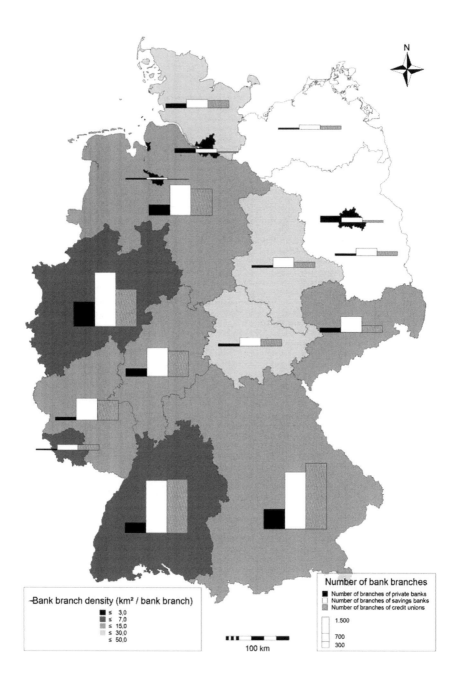

Figure 7.2   Bank branch densities in 2003
Source: Deutsche Bundesbank, 2004.

In this context, personal trust between the actors played a vital role (Pollard, 2003; Ennew and Martin, 1997). In the course of the internationalization of the banking industry, increased competition forced the banks to reorganize and standardize their processes in order to be able to work more cost-efficiently. Moreover, standardization has also become necessary due to political regulation. The Basel Accord, which defines capital requirements for banks, shall help to establish a more stable international financial system. It goes along with standardized credit rating systems which have increasingly replaced the traditional system of relationship banking.

Financing Renewable Energies-Focused SMEs in Rural Regions

The renewable energies sector in Germany has expanded rapidly. Between 2003 and 2013, the turnover in the industry rose by 334 per cent from 10.0 to 33.4 billion euros. Similarly, the number of employees climbed from 160,500 in 2004 to 377,800 in 2012. The highest employment ratios existed in Saxony-Anhalt, Brandenburg, and Mecklenburg-Western Pomerania. In 2012, 26.3 people out of 1,000 employees worked in the renewable energy industry in Saxony-Anhalt. In Brandenburg, this number amounted to 21.4, in Mecklenburg-Western Pomerania it was at 19.2. On average, 9.9 people out of 1,000 employees worked in the renewable energies industry in Germany (GWS, 2013).

Concerning the financing of businesses in the renewable energy industry, governmental support programmes play an important role. However, these mostly depend on the cooperation of a business with a credit institution, through which the application for and the payout of the funds are processed. Furthermore, the businesses themselves have to provide certain amounts of capital which are often partly financed by the serving credit institutions. Thus, good cooperation between renewable energies-focused SME and their house banks is crucial.

When applying for a bank credit, the SME management (SME-M) gives all rating-relevant information to the bank's customer adviser (BCA) who screens it and then hands it on to the back-office credit department. The application is analysed, and the decision whether a bank hands out a credit or not is made there in the back-office. Usually, there is no direct contact between the SME and the credit department (Handke, 2011). The customer adviser acts as a broker between the SME and the credit department and plays a key role in the communication process.

The bank's customer adviser is responsible for collecting the information the credit department needs. Therefore, this person decides whether a credit application should be considered or not. So the communication process between the SME management and the bank's customer adviser is crucial. In rural areas, where the distances between those two parties tend to be longer, the costs of collecting the information can be higher, too, which makes it more unattractive for bank's customer advisers to advocate credit applications.

Several studies argue that – due to this reason – the access to finance differs spatially according to the spatial organization of the financial system (Gärtner,

2010; Pieper, 2005; Leyshon and Thrift, 1995). That is to say, SMEs in rural areas, where the bank branch density, referring to the area per bank branch, is lower, are disadvantaged concerning their finance. Due to the processes of demographic shrinking the population density in rural areas in Germany is expected to decline further in the future (Gans, 2011; Bertelsmann-Stiftung, 2011). At the same time politics are pressurized to reduce subsidies for rural development in favour of growth strategies for metropolitan regions (BMVBW/BBR 2005, Tutt, 2007). In the context of providing services of general interest and ensuring equivalent living conditions in rural areas (BBSR, 2013; Rosenfeld, 2009; Kersten, 2009), the issue of financial exclusion has been raised (Leyshon and Thrift, 1995; European Commission, 2008).

However, in this chapter I argue that this is not necessarily the case. Instead, I argue that access to finance for SMEs is not a matter of geographical closeness but of social nearness between the business partners and very much depends on the relationship between the SME management and the bank's customer adviser. And here, personal trust still plays a vital role.

Theoretical Framework and Methodology

Intenseness of Interaction

In order to analyse the relationships between banks and renewable energies-focused SMEs in Germany, a theoretical framework was developed (Panzer-Krause, 2011). In the first part of this framework standardization and individualization are recognized as two extremes in a continuum of modes of interaction in bank–business relationships. Standardization as an anonymous mode of interaction allows – due to economies of scale – a high number of transactions at low costs. The price is defined by supply and demand, and there is a high transparency of costs. However, the loyalty of clients is relatively low. Individualization on the other hand is characterized by personal and custom-made elements of interaction. Despite being more costly, individualized bank–business relationships allow very good access to information for the bank and often go along with stronger loyalty from the clients. Therefore, individualized bank–business relationships correspond to the traditional form of relationship banking.

The intenseness of interaction reflects the overall degree of standardization and individualization in a bank–business relationship. It is differentiated between three levels of intenseness of interaction: low, medium, and high. The intenseness of interaction is a weighted, additive index which is composed of three indicators that reflect both individual perceptions and objective measures. The first indicator expresses the degree of individuality from the business' management's point of view. The SME-management evaluates to what extent the business relationship to their bank's customer adviser is on a personal basis and integrates custom-made elements. For this, a scale with discrete values from 0 (very standardized) to 3 (very individualized) is used. The second indicator measures the degree of individuality from the bank's customer adviser's point of view. The same scale as for the first indicator is employed. With the help of the third indicator the frequency of face-to-face contacts between the two business partners is evaluated. Here, a scale with discrete values from 0 (low frequency, maximally once a year) to 2 (high frequency, more than four times a year) is applied. This third indicator is part of the index, as face-to-face meetings especially have a negative influence on transaction costs.

The three values are added to an index which can take discrete overall values between 0 and 8. It must be pointed out that due to the smaller range of the third indicator, the weighting that is given to this indicator only lies at 25 per cent. In contrast, the weightings that are given to the first and the second indicator lie at 37.5 per cent. Therewith, the emphasis is put on the individuality in the bank–business relationships from the actors' point of view.

If the overall value lies between 8 and 6, the intenseness of interaction is considered high. An overall value between 5 and 3 represents a medium level of intenseness of interaction. Finally, a low intenseness of interaction is reached if the overall index value is between 0 and 2.

Modelling Trust

In the second part of the theoretical framework a model for the empirical ascertainment of trust was created (Panzer-Krause, 2011). The aim is to analyse to what extent trust between the two business partners can have an impact on the intenseness of interaction.

Luhmann (1989) recognizes trust as a social mechanism to reduce uncertainty. Thereby, many authors differentiate between systemic trust and personal trust (e.g. Luhmann, 1989; Coleman, 1990; Dibben, 2000; Glückler, 2004). However, only the latter is of interest in the context of this chapter since it refers to social relationships and represents an informal way of economic exchange. Trust is seen as situational on the one hand, but also as process-orientated on the other hand as it can change over time. Therefore, this chapter uses a model that includes five forms of trust which can evolve at different stages of a relationship. These five forms of trust can be put into a consecutive order concerning their quality: faith-based trust (A), dependence-based trust (B), competence trust (C), good-will trust (D), and confidence-based trust (E) (Figure 7.3).

Faith-based trust can develop at the beginning of a business relationship in an initial phase of becoming acquainted with someone. It can arise in situations which require interpersonal trust, but in which trust cannot be given on the basis of positive experiences in the past but only on the basis of faith. This form of trust only holds a low quality, as it is very unlikely that a superior form of trust is developed in the case that one of the actor's faith-based trust is breached. Dependence-based trust can follow a phase of positive early interaction. It is characterized by a mutual calculative assessment of the business partners.

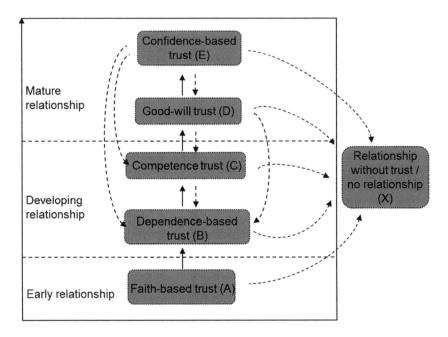

Figure 7.3   Model of situational trust
Source: Panzer-Krause, 2011.

Risks and possible benefits, which can result from the bank–business relationship, are weighed carefully by the two actors. Competence trust, which can be described as trust in the partner's competences and qualifications, has a higher quality than dependence-based trust. Smaller disappointments mostly do not lead to an ending of the business relationship. Good-will trust refers to the trust that the business partner abstains from opportunistic behaviour. It only tends to evolve in mature relationships and it is even more robust than competence trust. Confidence-based trust is seen as a form of trust that is characterized by a very high quality. It requires a clear identification with the partner and a long history of interaction.

In this chapter, the form of trust in a bank–business relationship is expressed in the following trust specification code:

[SME-M's form of trust] [reciprocity shown as arrows] [BCA's form of trust]

In order to reflect the process-oriented character of trust, I evaluate the forms of situational trust in bank–business relationships throughout the relationships. By means of this dynamic character of the analysis, the genesis of trust can be reconstructed.

Methodological Approach

On the basis of the theoretical framework, data collection was carried out in 2007 and 2008. Here, a sample of 20 relationships between banks and renewable energies-focused SMEs in Germany and their development over the years were analysed. For this purpose, interviews with both the SME managements and – where possible – their house banks' customer advisers were conducted, and the effects of different modes of interaction in relationships between banks and renewable energies-focused SMEs were analysed. Thereby, aspects of personal trust and its genesis in the context of geographical distance were the focal points. The analysed SMEs were spread all over Germany and worked in the following sectors of the renewable energies industry: solar energy (photovoltaic and solar heat), wind energy, hydraulic power, and biogas. The distance between the SMEs and their house banks varied between 0.5 and 190 km. A mix of quantitative and qualitative methods was applied:

1. The intenseness of interaction was identified.
2. The forms of trust were evaluated and the genesis of mutual trust was reconstructed in its context for all relationships analysed.

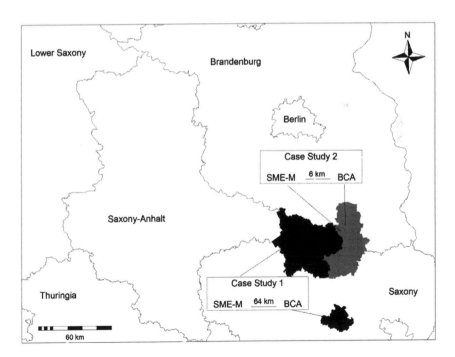

Figure 7.4   Overview of the two case studies

In order to identify adequate cases for the study, SMEs working in the renewable energies sector were invited to participate. For those SMEs that agreed, the next step was to contact their house bank's customer adviser encourage them to collaborate. As banking practice regarding privacy forbids bank employees to reveal information about their clients to third parties by law, the SME managements explicitly had to release their bank's customer adviser from this banking practice. This complicated the search for adequate cases. However, in 14 cases interviews with both the SME management and their bank's customer adviser could be arranged. In six cases, interviews could only be realized with the SME management. Different reasons such as the SME's lack of knowledge about who their bank's customer adviser was or the SME's reluctance to involve the bank's customer adviser prevented interviews with the SME's bank's customer advisers.

In this chapter, I explicitly focus on two case studies in which two renewable energies-focused SMEs located in sparsely populated rural areas in Brandenburg as well as their relationships to their house banks are analysed (Figure 7.4). The intenseness of interaction and the forms of trust are investigated and how these correspond to each other is analysed. Then which mode of interaction is preferable in bank–business relationships is discussed.

The two case studies were selected because they represent quite contrary surrounding conditions. While Case Study 1 is characterized by a relatively long geographical distance between the SME location and its house bank (64 km), in Case Study 2 the geographical distance between the two actors is comparatively small (6 km). Nevertheless, the case studies reveal that it is not the geographical closeness that ensures access to finance but rather a higher quality of trust combined with a medium intenseness of interaction. While the majority of the cases of the empirical study were characterized by a high intenseness of interaction, Case Study 1 especially illustrates that transaction costs can be reduced if the intenseness of interaction is reduced. Thus, the risk of financial exclusion can be minimized. However, the aim of the study was not to generalize the results but rather to analyse what influence personal trust between the actors may have concerning different forms of interaction.

Both renewable energies-focused SMEs analysed still operate today and thus show a continuity of their economic activity in rural areas and demonstrate the efficacy of their particular form of financing. Therefore, they are suitable examples in order to illustrate financial aspects of supply chains and the role renewable energies-focused SMEs play in rural development.

Results

Case Study 1: Financing Biogas Plants in Brandenburg

The first case study deals with a company whose head office is located in the county Elbe-Elster in Brandenburg, a region that is classified as a sparsely populated rural area in Germany by BBSR (2011). The company runs four biogas plants in rural Brandenburg and Saxony-Anhalt which produce electricity from organic waste and animal by-products. It also develops procedures for the preparation of organic waste products for fermentation. Primary agricultural products such as maize are not used.

The company's management initiated and planned the biogas plants, monitored their production process and the erection of the plants and has run them since their completion. At the time of the data collection, three biogas plants were completed; the fourth was in the planning process. In 2007 the business had 55 employees and generated a turnover of about six million euros.

In order to run the biogas plants efficiently, the business depends on a network of suppliers such as supermarkets or slaughterhouses that ensure a continuous replenishment of organic waste and animal by-products. Therefore, long-term contracts with regional suppliers were a prerequisite for the financing of the biogas plants as well as for setting up credit lines in order to ensure the firm's daily liquidity.

The SME management worked together with a large, nationally operating bank which had its closest bank branch in Dresden from where it serviced the business. The geographical distance between the company's head office and the location of the bank's customer adviser was 64 km. Approximately 80 per cent of the financing was accomplished by long-term loans. The first project financing was initiated in the house bank's branch in Munich; when a new investor took the company over in 2005, the financing was handed over to the bank's branch in Dresden. The company's general manager was employed by the proprietor. The business's house bank served corporate clients with an annual turnover of more than three million euros onsite, meaning that the bank's customer adviser visited the SME management at the company's location for business meetings. The bank's business adviser specialized in the engineering and renewable energies market and served about 50 clients.

The intenseness of interaction between the SME management and its bank's customer adviser scored a value of 5 (indicator 1 = 2; indicator 2 = 2; indicator 3 = 1) and was therefore at a medium level since its establishment in 2005. The SME management as well as the bank's customer adviser characterized the business relationship as very positive while not very personal, but rather personal. Each year, a maximum of two face-to-face meetings between the two partners were conducted. All other matters were handled via telephone or email. For this, information was not only exchanged between the SME management and the customer adviser but also between the SME administration staff and various specialists in the bank.

Concerning the role of trust in the relationship between the two partners, the bank's customer adviser clarified that 'trust in the bank–business relationship is a basis, but it is not the essential criteria for granting a loan'[1] (BCA CS-1[2]).

---

1 Original quotation: 'Vertrauen in der Kundenbeziehung ist eine Grundlage, aber es ist nicht das entscheidende Kriterium für eine Kreditausreichung' (BCA CS-1).

2 CS = case study.

The bank's customer adviser stressed that personal trust, however, makes a big contribution to organize specific situations in a time-efficient way. In this way, firms can benefit from economies of time. For both partners, personal trust in the competences and qualifications of the business partner seemed to be crucial. The bank's customer adviser pointed out: 'I think it is important to be well versed in the company's business area, so we speak the same language'[3] (BCA CS-1). In this context the bank's customer adviser also made it clear that the SME management had 'a very good expertise for the whole biogas market, so we can talk about not only this single project but also about other things all around. [The SME-M] is a contact partner for me for such business dealings'[4] (BCA CS-1).

Similarly, the SME management held the view that in a bank–business relationship professional expertise is more important than the social side of the relationship. He pointed out that according to his experiences, regional banks often cannot provide professional expertise. From his point of view, the bank's customer adviser was factual and interested, which were for him the most important things. The trust specification code for this bank–business relationship can be characterized as $C \rightleftharpoons C$ after it went through lower qualities of trust at the very beginning of the business relationship. At the same time, the SME management emphasized that a bigger geographical distance between the actors, like the 64 km that parted the two business partners, was not an obstacle for the development of personal trust: 'You can have a good relation to someone, even though you do not meet him every day'[5] (SME-M CS-1).

The bank's customer adviser emphasized that face-to-face meetings often took place following recent occurrences: 'When there is something that needs to be dealt with then there are phases in which we meet more often, and then afterwards again on an irregular basis'[6] (BCA CS-1).

The SME management added that about two face-to-face meetings are sufficient in order to clarify things: 'When it is about a project, then maybe two meetings, then everything is clear'[7] (SME-M CS-1).

When there are no special occurrences, the business partners only intended to meet once a year in order to keep each other informed and to go through potential

---

3   Original quotation: 'Ich denke, es ist wichtig, dass man sich auskennt, dass man die gleiche Sprache spricht' (BCA CS-1).

4   Original quotation: [Die KMU-Geschäftsleitung hat ein] 'sehr gutes fachliches Verständnis, auch für den gesamten Markt im Biogasbereich, wo man sich nicht nur über das einzelne Projekt unterhalten kann, sondern auch über die Sachen ringsherum. Und für mich ein Ansprechpartner auch für solche Geschäfte' (BCA CS-1).

5   Original quotation: 'Sie können trotzdem zu jemandem einen guten Kontakt haben, auch wenn Sie ihn nicht jeden Tag sehen' (SME-M CS-1).

6   Original quotation: '[…] wenn es etwas zu erledigen gibt, dann gibt es Phasen, wo man sich häufiger trifft, und dann ist das wieder in unregelmäßigen Abständen' (BCA CA-1).

7   Original quotation: 'Wenn es um ein Projekt geht, vielleicht zweimal Treffen, dann ist das klar' (SME-M CS1).

future financing plans. Such meetings took place at the business's location. The bank's customer adviser noted: 'Concerning cooperate banking, I would say, the bank's customer adviser needs to be mobile. Not the business client comes to the bank, the way is vice versa. That's why it is not crucial where the bank has its location [...]'[8] (BCA CS-1).

In summary it can be said, that the company was characterized by the fact that both the investors and the business manager were not locals but external actors that came from other regions. Financing for the first biogas plant was brought into the rural county of Elbe-Elster and thus helped to create numerous jobs within the region. After a short phase of becoming acquainted with each other, the personal trust between the SME management and the bank's customer adviser based in Dresden climbed for both business partners from faith-based trust to dependence-based trust and then to competence trust where it stagnated. This form of mutual trust is of higher quality, although it is not as robust as good-will trust. Nevertheless, it developed despite the fact that the business partners were parted by quite a big geographical distance of 64 km.

At the same time it has to be stressed, that the business's house bank only serviced the business onsite because it had an annual turnover of more than three million euros. Thus, the proportion between transaction costs and the potential yield for the bank was attractive enough. Businesses with a smaller annual turnover might suffer from financial exclusion. Altogether, the mode of interaction in this bank–business relationship can be identified as optimal as it was characterized by a medium intenseness of interaction and mutual competence trust. In fact the relatively high quality of trust enabled the partners to work on the basis of a medium intenseness of interaction. The geographical distance of 64 km between the SME and its house bank was not a disadvantage for the business's financing.

Case Study 2: Financing the Production of Solar Panels and Solar Heat Panels in Brandenburg

The second case study focuses on the bank–business relationship between an SME working in the solar panel and solar heat panel industry and its house bank. The business is located in the sparsely populated rural county of Oberspreewald-Lausitz in Brandenburg. The company was founded in 1997 and produces solar panels and solar heat panels with a focus on solar heat panels. It is an affiliate of a local glass producer. The initial crisis of the solar market forced the company to reduce its number of employees from 22 to 16 in 2008. The turnover totalled 1.7 million euros in 2007. In 2006 it was 2.5 million euros, so that there had been

---

8   Original quotation: 'Für das Firmenkundengeschäft, da würde ich sagen, da muss der Firmenkundenbetreuer mobil sein und möglichst beim Unternehmen, weil, der Firmenkunde kommt nicht in die Bank, sondern der Weg ist andersherum. Deswegen ist es nicht so entscheidend, wo die Bank ihren Sitz hat [...]' (BCA CS-1).

a sales collapse of about 30 per cent: 'We have ups and downs, because the solar industry is really very, very unstable'[9] (SME-M CS-2).

While the company sourced parts for its production from various Asian suppliers, 85 per cent of the products were sold on the national markets and 15 per cent of the products were sold internationally to France, Poland, and Tunisia. When engaging in bigger projects, business partners of the supply chain often inquired about the company's rating score from its house bank '[...] in order to make sure: How well is the company going?'[10] (SME-M CS-2).

The business' house bank was located just 6 km from the company's location. It is a regional bank operating solely in the area. The bank's customer adviser serviced about 150 clients and was not focused on certain sectors. The bank's customer adviser had worked with the company for seven years at the time of data collection. The business did not have any other bank connections. It had taken out several loans from the bank. For projects usually a mix of financing was used; 40 per cent were financed by loans, 40 per cent were financed by subsidies of the federal state of Brandenburg and 20 per cent were self-financed.

The intenseness of interaction between the SME management and the bank's customer adviser could be valued as high at the time of data collection in 2008 as the overall index value scored 6 (indicator 1 = 2; indicator 2 = 2; indicator 3 = 2). Both business partners characterized the business relationship – equally to Case Study 1 – as rather personal. However, while the bank's customer adviser found the overall relationship very positive, the SME management described it only as rather positive. Thereby, face-to-face meetings were conducted on a much more frequent basis than in Case Study 1. The business partners arranged between 8 and 10 formal meetings each year.

Concerning the form of mutual trust, it can be stated that both the SME management and the bank's customer adviser broached the issue but illustrated that they acted on a very calculative basis. The SME management declared:

> The bankers need to be able to expect that I give them the information they want outright. If there are critical situations, then some evidence should be handed over in the forefront so that they are not thrown in at the deep end. That is the kind of trust that I have to give my bank. On the other hand I try to minimize the information the bank gets – only what is necessary. The bank would like to see clearly more.[11] (SME-M CS-2)

---

9   Original quotation: 'Es gibt hier Aufs und Abs, weil die Branche ganz einfach auch wirklich sehr, sehr wackelig ist' (SME-M CS-2).

10   Original quotation: '[...] wird diese Ratingzahl eingeholt, um sich Sicherheit zu verschaffen: wie läuft denn das Unternehmen überhaupt' (SME-M CS-2).

11   Original quotation: 'Die Banker müssen von mir erwarten können, dass ich ihnen die Informationen, die sie haben wollen, offen und ehrlich rüberreiche. Wenn es zu Krisensituationen kommt, dass man dann eben bewusst im Vorfeld schon bestimmte Hinweise rübergegeben hat, dass sie das nicht kalt überfällt. Das ist das Vertrauen, das ich

He added that his house bank worked in the same way:

> By the way, what the bank does is not any different. The kind of information we really claim, about the possibilities there are etc. – this is what we get told. But if we don't approach the bankers actively, we do not come to know the important things that could help the company. Even our house bank that has been working with us for the last 15 years does not let us know more than what we really claim.[12] (SME-M CS-2)

Similarly the bank's customer adviser made clear that he centred the bank–business relationship around the potential yield for the bank by always considering 'how closely the bank and the client work together, how closely are we interlocked via loans accounts etc.?'[13] (BCA CS-2). What this form of trust means on a working basis is well illustrated by the following statement of the SME management:

> In certain situations we tend to make use of blackmailing strategies. Thus, we force our house bank to think about certain things. But we do not do it in a way that the other side could lose face. Because on the other hand we do not want to be labelled in a certain way in case our company is not doing so well – you never know what might happen in the next years. So far it has always been a relatively fair cooperation.[14] (SME-M CS-2)

Although both business partners referred to their partner's competence, this bank–business relationship was driven by calculative aspects. Therefore, the bank–business relationship was characterized by mutual dependence-based trust; the trust specification code was $B \rightleftharpoons B$.

---

meiner Bank entgegen bringen muss. Auf der anderen Seite versuche ich die Fakten, die die Bank bekommt, wirklich zu minimieren – das, was notwendig ist. Die Bank würde gern deutlich mehr sehen' (SME-M CS-2).

12 Original quotation: 'Anders ist es bei der Bank im Übrigen auch nicht. Was wir tatsächlich einfordern an Wissen, welche Möglichkeiten es gibt usw., das wird uns gewährt. Wenn wir aber nicht selbständig auf die Banker zugehen, erfahren wir auch wesentliche Dinge, die der Firma helfen könnten, nicht von den Bankern. Auch von unserer seit 15 Jahren begleitenden Hausbank erfahren wir nicht mehr' (SME-M CS-2).

13 Original quotation: '[...] wie eng arbeiten tatsächlich auf der Geschäftsebene die Bank und der Kunde zusammen? Wie stark sind wir miteinander verquickt – über Darlehenkonten usw.?' (BCA CS-2).

14 Original quotation: 'In bestimmten Situationen neigen wir auch zu erpresserischen Handlungen. Wir zwingen auch unsere Hausbank dazu, über bestimmte Dinge nachzudenken. Aber wir machen es eben nicht so, dass wir die Leute in einer Form erpressen, wo die andere Seite vielleicht das Gesicht verliert. Denn wir wollen auf der anderen Seite, wenn es uns vielleicht mal nicht so gut geht – man weiß ja nicht, wie es sich in den nächsten Jahren alles gestaltet – wollen wir genauso nicht in die Ecke gestellt werden. Und bisher war das immer ein relativ faires Miteinander' (SME-M CS-2).

For the business partners in Case Study 2, face-to-face communications were much more important compared to the actors in Case Study 1. These face-to-face contacts were mostly initiated by the SME management: 'I have a few principals concerning my business relationship to the bank, and one is that I hand in the important documents personally. Maybe [the bank's customer adviser] has questions concerning certain matters, then I can answer them straight away'[15] (SME-M CS-2).

Nevertheless, the bank's customer adviser also serviced the client at his business location depending on which way it was easier to organize: 'As the case may be. If [X] is hereabouts then he comes to see me, otherwise I go to him. However, he is away a lot [on business]'[16] (BCA CS-2).

But similarly to Case Study 1, this bank–business relationship was characterized by quiet phases and by phases in which more meetings were arranged: 'There are phases when urgent matters have to be solved – problems or whatever, so that we meet twice a week. There are also phases, in which there are no urgent matters. Then you don't hear from each other for two months'[17] (BCA CS-2).

The house bank was not the only bank showing interest in having the company as a corporate client: 'Even though this is our only bank connection, we have very intense contacts to [other large, nationally operating banks ...]. Of course, they try to woo you away'[18] (SME-M CS-2).

The bank–business relationship experienced a difficult time when the company went through a crisis in 2002. Although the form of mutual trust was not very robust, the business relationship survived this crisis:

> We experienced a crisis in our company in 2002. That was quite strenuous, but in such situations you see whether your partner helps you or not. Back then, we got a temporary financing [...]. This temporary financing has nearly been zeroized by now. It was not a huge sum, but it was clearly above 100,000 euros.[19] (SME-M CS-2)

---

15   Original quotation: 'Ich habe ein paar grundsätzliche Gedanken zu so einer Geschäftsbeziehung mit der Bank und dazu gehört u.a., dass ich die wichtigen Unterlagen immer persönlich reinreiche. Vielleicht hat [Kundenbetreuer] ja noch Fragen zu bestimmten Sachen, die kann ich ihm dann gleich beantworten' (SME-M CS-2).

16   Original quotation: 'Je nachdem, wie es sich anbietet. Wenn [X] in der Gegend ist, kommt er hier lang, ansonsten fahre ich vorbei. Er ist aber recht viel [beruflich] unterwegs' (BCA CS-2).

17   Original quotation: 'Es gibt akute Phasen, wenn ein Problem zu lösen ist oder – wie auch immer – irgendetwas anliegt, dass man sich zweimal pro Woche sieht. Es gibt auch Phasen, wo gerade nicht wirklich akuter Bedarf besteht. Dann hört man zwei Monate nicht voneinander' (BCA CS-2).

18   Original quotation: 'Auch wenn wir mit der Bank als alleinige Bank arbeiten, haben wir doch auch [zu anderen Großbanken] sehr intensive Kontakte. [...] Natürlich versuchen, die einen auch abzuwerben' (SME-M CS-2).

19   Original quotation: 'Wir haben hier also auch das Krisenszenario live im Betrieb 2002 erleben dürfen. Das war für uns relativ anstrengend, aber in solchen Situationen sieht

Possibly the house bank backed this crisis up as it works in a rural region which is characterized by a difficult economic situation and where comparable firms like the business partner and its parent company are scarce. And knowing that other banks intended to entice the client might have induced the house bank to help the business partner. Again, this acting shows a calculative approach. However, the bank's customer adviser was not willing to give any detailed information about this incident.

Altogether, it can be said that the company had its ups and downs due to the problems in the international solar panel and solar heat panel market. Having a parent company was clearly an advantage concerning the securing of finance. Despite the fact, that the intenseness of interaction was high, the business relationship was only characterized by mutual dependence-based trust which surprisingly survived a bigger crisis in the company. However, due to the numerous face-to-face meetings throughout the year, this mode of interaction depends on short geographical distances between the business partners and only works well as long as the regional bank's branch is close-by. Especially from the bank's point of view relatively high transaction costs were incurred, which reduce the potential yield. Therefore, the mode of interaction can be considered as less efficient than in Case Study 1.

Conclusion

It can be concluded that renewable energies-focused SMEs in rural regions have access to finance and it does not primarily depend on their geographical location. Instead, access to financing possibilities depends on the transaction costs incurred for banks. These transaction costs are higher when the intenseness of interaction is high. In this case there is a potential for financial exclusion. However, personal trust can have a positive influence on the intenseness of interaction. Especially, more robust forms of trust like competence trust have the potential to reduce a high intenseness of interaction to a medium level without affecting the reduction of information asymmetries.

Concerning the genesis of trust, it can be stated that personal trust mostly builds up in phases in which the need for face-to-face contact arises. In times with low or no requirements for face-to-face contacts the bank–business relationship can live off the trust evolved beforehand. These are the phases which provide a high potential for savings for the credit institutions.

As Case Study 1 showed, economic exchange can be conducted without the actors' permanent geographical proximity. Spatial patterns are instead massively

---

man ja, ob einem geholfen wird oder nicht. Und wir haben damals eine Zwischenfinanzierung aufgenommen, also, um nicht in den Kontokorrent gehen zu müssen, haben wir ganz einfach auch noch mal eine gute Kreditlinie zusätzlich ausgehandelt. Die ist mittlerweile – ich sage mal – fast auf Null gefahren. Also, das war kein Riesen Betrag, aber der war schon deutlich höher als 100.000 Euro' (SME-M CS-2).

influenced by social aspects. Therefore, the chapter demonstrates that renewable energies-focused SMEs located in rural regions in Germany are not per se disadvantaged concerning their finance opportunities. On the contrary, Case Study 2 made clear that a high intenseness of interaction does not necessarily lead to more robust forms of trust.

The chapter shows that nationally operating large banks which do not run bank branches in rural areas anymore still service companies located in these regions. Nevertheless, the relationship between the bank and the business can be organized efficiently if the intenseness of interaction can be reduced to a medium level.

The results make clear that renewable energies-focused SMEs in rural areas can contribute to a positive development of those regions by creating jobs and thus raise the acceptance of renewable energy production sites. Therefore, renewable energies have the potential to be the engine for rural development. The financing of certain parts within the supply chain in the renewable energies industry does not depend on geographical closeness between credit institutions and renewable energies-focused SMEs but rather on their social nearness. However, further research on the spatial dimension of financial supply chains remains to be conducted.

Concerning policy implications, the findings suggest that on the national level there is no need for policy interventions, i.e. support for decentralized structures of regional banks. However, on the local and regional level, politics can optimize the economic promotion of rural areas by not exclusively focusing their activities on the regional context itself, for example by insisting on creating regional economic cycles. Rather they should stress the importance of rural–urban interconnections and integrate their activities in cross-regional and supraregional settings and networks. However regional development agencies often underestimate the relevance of such rural–urban settings and networks.

References

Bair, J., Gibbon, P., and Ponte, S., 2008. Governing Global Value Chains. An Introduction. *Economy and Society* 37, 315–38.

BBSR, 2013. *European Atlas of Services of General Interest*. Bonn: BBR.

———— 2011. Laufende Raumbeobachtung – Raumabgrenzungen. Siedlungsstrukturelle Kreistypen [online]. Available at: http://www.bbsr.bund.de/nn_1067638/BBSR/DE/Raumbeobachtung/Raumabgrenzungen/Kreistypen4/kreistypen.html [accessed: 15 November 2013].

Bertelsmann-Stiftung, 2011. Deutschland im demographischen Wandel 2030. Datenreport. Bielefeld: Bertelsmann-Stiftung.

BMU, 2011. Das Energiekonzept der Bundesregierung 2010 und die Energiewende 2011 [online]. Available at: http://www.bmu.de/fileadmin/bmu-import/files/pdfs/allgemein/application/pdf/energiekonzept_bundesregierung.pdf [accessed: 15 November 2013].

BMU and BDI, 2013. *Green Economy in der Praxis* [Erfolgsbeispiele aus der Praxis]. Paderborn: BMU/BDI.

BMVBW/BBR, 2005. Öffentliche *Daseinsvorsorge und demographischer Wandel.* Berlin/Bonn: BMVBW/BBR.

Clark, G.L. and Wójcik, D., 2007. *The Geography of Finance: Corporate Governance in a Global Market-Place.* Oxford: University of Oxford Press.

Coleman, M., 1990. *Foundations of Social Theory.* Cambridge: Harvard University Press.

Deutsche Bundesbank, 2013. Entwicklung des Bankstellennetzes im Jahr 2012. Frankfurt/Main: Deutsche Bundesbank.

——— 2012. Entwicklung des Bankstellennetzes im Jahr 2011. Frankfurt/Main: Deutsche Bundesbank.

——— 2008. Entwicklung des Bankstellennetzes im Jahr 2007. Frankfurt/Main: Deutsche Bundesbank.

——— 2004. Entwicklung des Bankstellennetzes im Jahr 2003. Frankfurt/Main: Deutsche Bundesbank.

——— 2000. Entwicklung des Bankstellennetzes im Jahr 1999. Frankfurt/Main: Deutsche Bundesbank.

Dibben, M., 2000. *Exploring Interpersonal Trust in the Entrepreneural Venture.* Basingstoke: Macmillan Press.

Ennew, C. and Martin, R., 1997. Smaller Businesses and Relationship Banking: The Impact of Participative Behavior. *Entrepreneurship: Theory & Practice* 21 (4), 83–93.

European Commission (ed.), 2008. Financial Services Provision and Prevention of Financial Exclusion. Brussels: European Commission.

French, S., Leyshon, A., and Wainwright, T., 2011. Financializing Space, Spacing Financialization. *Progress in Human Geography* 35 (6), 798–819.

Gans, P., 2011. *Bevölkerung. Entwicklung und Demographie unserer Gesellschaft.* Darmstadt: Primus.

Gärtner, S., 2010. Die räumliche Dimension im Bankgeschäft: Regionale Finanzintermediäre in strukturschwachen Räumen. In: Christians, U. and Hempel, K. (eds), *Unternehmensfinanzierung und Region: Finanzierungsprobleme mittelständischer Unternehmen und Bankpolitik in peripheren Wirtschaftsräumen.* Hamburg: Verlag Dr. Kovac, pp. 205–30.

Gereffi, G., Humphrey, J., and Sturgeon, T., 2005. The Governance of Global Value Chains. *Review of International Political Economy* 12 (1), 78–104.

Glückler, J., 2004. *Reputationsnetze. Zur Internationalisierung von Unternehmensberatern. Eine relationale Theorie.* Bielefeld: Transcript.

GWS, 2013. *Erneuerbar beschäftigt in den Bundesländern: Bericht zur aktualisierten Abschätzung der Brutto-Beschäftigung 2012 in den Bundesländern.* Osnabrück: GWS.

Handke, M., 2011. *Die Hausbankbeziehung. Institutionalisierte Finanzierungslösungen für kleine und mittlere Unternehmen in räumlicher Perspektive.* Münster: LIT.

Kersten, J., 2009. Wandel der Daseinsvorsorge. Von der Gleichwertigkeit der Lebensverhältnisse zur wirtschaftlichen, sozialen und territorialen Kohäsion. In: Neu, C. (ed.), *Daseinsvorsorge. Eine gesellschaftswissenschaftliche Annäherung.* Wiesbaden: VS Verlag für Sozialwissenschaften, pp. 22–38.

Klagge, B., 2013. Governance-Prozesse für erneuerbare Energien – Akteure, Koordinations- und Steuerungsstrukturen. In: Arbach, C. and Klagge, B. (eds), *Governance-Prozesse für erneuerbare Energien.* Hannover: Akademie für Raumforschung und Landesplanung, pp. 7–16.

———— 2010. Das deutsche Banken- und Finanzsystem im Spannungsfeld von internationalen Finanzmärkten und regionaler Orientierung. In: Kulke, E. (ed.), *Wirtschaftsgeographie Deutschlands.* Heidelberg: Spektrum Akademischer Verlag, pp. 287–302.

Leyshon, A. and Thrift, N., 1995. Geographies of Financial Exclusion: Financial Abandonment in Britain and the United States. *Transactions of the Institute of British Geographers* 20, 312–41.

Luhmann, N., 1989. *Vertrauen. Ein Mechanismus zur Reduktion sozialer Komplexität.* Stuttgart: Enke.

Marshall, J.N., Pollard, J., and Wray, F., 2010. Finance and Local Regional Economic Development. In: Pike, A., Rodríguez-Pose, A., and Tomaney, J., (eds), *A Handbook of Local and Regional Development.* London: Routledge, pp. 356–70.

Musil, R. and Zademach, H.-M., 2012. Global Integration Along Historic Pathways. Vienna and Munich in the Changing Financial Geography of Europe. *European Urban and Regional Studies* 2012, 1–21.

Panzer, S., 2008. Rural Development and Social Embeddedness: Banks and Businesses in Thuringia, Germany. In: Tamásy, C. and Taylor, M. (eds), *Globalising Worlds and New Economic Configurations.* Farnham/Burlington: Ashgate, pp. 197–208.

Panzer-Krause, S., 2011. *Ohne Moos nichts los – KMU-Finanzierungen unter den Bedingungen sich wandelnder Finanzmärkte. Eine Untersuchung von Beziehungen zwischen Banken und KMU.* Hamburg: Verlag Dr. Kovac.

Pieper, C., 2005. *Banken im Umbruch. Strukturwandel im deutschen Bankensektor und regionalwirtschaftliche Implikationen.* Münster: LIT.

Pike, A. and Pollard, J.S., 2010. Economic Geographies of Financialization. *Economic Geography* 86 (1), 29–51.

Pohl, M., 2013. Regionalwirtschaftliche Bedeutung der Windenergie in Nordwestdeutschland – ein wichtiger Aspekt von Planungs – und Governance-Prozessen. In: Arbach, C. and Klagge, B. (eds), *Governance-Prozesse für erneuerbare Energien.* Hannover: Akademie für Raumforschung und Landesplanung, pp. 17–30.

Pollard, J.S., 2003. Small Firm Finance and Economic Geography. *Journal of Economic Geography* 3 (4), 429–52.

Rosenfeld, M.T.W., 2009. 'Gleichwertigkeit der Lebensverhältnisse' zwischen Politik und Marktmechanismus. Zusammenfassende Bewertung der Befunde und Schlussfolgerungen für regionale Entwicklungsstrategien. In: Rosenfeld, M.T.W. and Weiß, D. (eds), *Gleichwertigkeit der Lebensverhältnisse zwischen Politik und Marktmechanismus. Empirische Befunde aus den Ländern Sachsen, Sachsen-Anhalt und Thüringen*. Hannover: Akademie für Raumforschung und Landesplanung, pp. 253–8.

Sassen, S., 1991. *The Global City: New York, London, Tokyo*. Princeton: Princeton University Press.

Tutt, C., 2007. *Das große Schrumpfen*. Berlin: Berlin Verlag.

Wotha, B., 2013. Planerische Möglichkeiten zur Steuerung der Standortentwicklung und Verbesserung der Akzeptanz von Biogasanlagen. In: Arbach, C. and Klagge, B. (eds), *Governance-Prozesse für erneuerbare Energien*. Hannover: Akademie für Raumforschung und Landesplanung, pp. 69–93.

Chapter 8

Securing Local Supply in Rural Areas: The Role of Wholesale Cooperations in Central Hesse, Germany

Anika Trebbin, Martin Franz, and Markus Hassler

Introduction

Retailing in rural areas is undergoing dynamic transformations: individual shops' sales areas keep increasing in size while shop locations become more and more concentrated. As a result, there are more and more areas where a classical supermarket or discounter can only be reached after a longer car drive. These developments suggest that areas without a food supply might develop in rural regions which can threaten the food security of their population, especially of those who are less mobile. Behind this trend are the strategies of large integrated retail and wholesale groups like Edeka, Rewe, and Aldi that focus on establishing ever larger shops while closing down smaller outlets.

This chapter examines whether new structures of local supply and patterns of trade organization emerge in rural areas beyond these trends of increasing sales area and spatial concentration of retail outlets. It also explores what preconditions need to exist for such new structures to emerge. In doing this, we focus particularly on supply options for small supermarkets that are below the size limit of stores Edeka or Rewe keep in their supply network. The area of study is Central Hesse in Germany. This region lies north of the Rhine Main Area and comprises of both rural and urban areas. The rural areas of this region have been hit particularly hard by the ramifications of demographic change and outflow of people. With this set of problems, Central Hesse is comparable to many other regions in Germany and Europe.

The core question and broader theme of this chapter – the local supply situation in rural areas – have been relatively neglected in academic literature and research in the past. The discussion about local supply services in Germany was most vibrant in the 1970s and 1980s (Kuhlicke et al., 2005). Since then, it has been focusing more on retail in urban areas and city centres. If local supply in rural areas is being examined, most studies look into alternative and pilot projects such as neighbourhood/ convenience stores and new concepts for mobility (Kuhlicke et al., 2005).

This chapter is based on 10 qualitative interviews with wholesale and retail companies, representatives of the Departments for Rural Areas in the five administrative districts of the study region, and the department for large-scale retail of the regional authority of Central Hesse. Additionally, a standardized survey was conducted covering 102 small supermarkets in Central Hesse run by independent retailers. Data collection took place between November 2009 and March 2010. In April 2011, another six retailers were interviewed by telephone. All interviews were processed through content analysis. The interviews were conducted in German. Quotes have been translated into English by the authors for this chapter.

In the next section we will conceptually embed this chapter's main question into the field of geographical research on trade. Following this, the retail situation in Germany's rural areas is outlined. Then, the case study region is analysed. This analysis shows that in rural Central Hesse, the structural change in retail has opened up space for new trade structures to evolve as a new basis for rural local supply. This new space is successfully occupied by medium-sized wholesalers to expand and ensure their business. The inhabitants of the areas concerned profit from this trend as easy-to-reach local supply options are being preserved.

Local Food Supply and Spiral Movement Theory

Easy access to goods for everyday consumption is being increasingly considered as problematic in recent years, especially in rural areas. Even in larger villages or neighbourhoods of larger cities, the inhabitants' local supply might no longer be secured (Kuhlicke et al., 2005). This could exacerbate people's access to fresh produce in particular and especially for those who are little or less mobile. In many European countries, such as the United Kingdom, Ireland, and Germany, the transition in retail formats and concentration processes in the retail trade have led to the disappearance of corner shops, village shops and neighbourhood shops (Furey et al., 2001; Kirby, 1987; Kuhlicke et al., 2005). As a result, there are now regions in which getting one's daily supply of everyday goods is no longer possible without a car or a well-developed public transport network.

But how did the transition in retail formats that underlies these problems occur? Mainly responsible is the concentration process that is happening because of the expansion of chain stores and the simultaneous decline of independent retailer businesses. This process is enhanced by economies of scale that lead to ever increasing sales areas and the centralization of supply chains. As a result, only stores in locations that have a large catchment area and therefore a large potential to generate turnover remain competitive (Kuhlicke et al., 2005).

In the 1990s a large number of studies examined this structural transformation in retail. Amongst them were studies referring to the German retail market (Heinritz, 1991; Kulke, 1992, 2006; Wortmann, 2003) as well as studies examining this phenomenon in an international context for other countries and regions (Bloch et al., 1994; Miller et al., 1999; Morganosky and Cude, 2000; Redondo, 1999). Most of the theoretical concepts used in these studies were developed between the

1960s and 1980s. They can be differentiated into the three broad conceptual strands of (1) environmental theories (for further details see Bartels, 1981; Globerman, 1978; Shaffer, 1973); (2) conflict theories (see for example Gist, 1968; Izraeli, 1971; Martenson, 1981); and (3) cyclical theories such as the so-called *retail accordion* (Hollander, 1966), the *retail wheel* (McNair and May, 1976), the *retail life cycle* (Davidson et al., 1967; Kulke, 1992) and the *polarization model* (Brown, 1987; Kirby, 1987). From the group of cyclical theories, the polarization model is particularly interesting for the question discussed here: '[…] the polarisation principle contends that the well-documented trend towards fewer but larger retail establishments is counterbalanced by a renaissance of the small shop sector' (Brown, 1987: 157).

Aspects from all three theory strands (environmental, conflict, and cyclical theories) are combined in Agergård et al.'s (1968) *spiral movement theory.* Referring only to urban areas, the spiral movement theory is based on the assumption that the living standard of the population in a given area is steadily improving which in turn leads to an increase in motorization. Because of the latter, store locations that are easily accessible by car or public transport are becoming more attractive and sales areas of retail outlets are increasing. This increase of the sales area and the catchment area of retail outlets is a gradual but constant process and finally, store locations become more and more spatially concentrated while distances between individual outlets grow and therefore the distance shoppers have to cover in order to reach the store (Agergård et al., 1968).

> With increasing car-ownership, the distance parameter, up to a certain level, is of decreasing importance […]. But at a certain time, the distance becomes so long that this parameter suddenly becomes important, as there is always a certain relationship between the attraction and the distance which the customer is willing to travel from his home. This reaction on the customer's side against travelling distance is, naturally, first observed within the convenience goods sector; this is due to the relatively small attraction of these articles. (Agergård et al., 1968: 15)

Finally, areas emerge in between those commercial centres that are large enough for new, smaller businesses to locate. As an example of this development Agergård et al. (1968) mention the evolution of neighbourhood shops in the United States. In Germany a similar development can be observed. Here, small grocery shops that sell fresh foods such as fruit and vegetables and that are often owned by citizens originating from Turkey or Arab countries have sprung up in recent years. 'As a reaction to the longer distances for customers, caused by the larger shops and centers, smaller shops are emerging with distance as their main parameter in the competition' (Agergård et al., 1968: 35).

Attempts to empirically verify Agergård et al.'s (1968) theory that smaller shops are emerging in between the larger shops have not been successful so far. Heinritz (1989) tried to verify the theory with a case study in Bavaria, Germany.

He concluded that although phenomena like the increase in sales area, the assortment and services in grocery stores could well be observed, such developments always go in hand with reverse trends. Thus a relatively simple model such as Agergård et al.'s theory could not be applied to the development of the retails sector in general (Heinritz, 1989).

However, the study conducted by Heinritz (1989) and Agergård et al.'s (1968) theory refer to urban areas. An attempt to transfer the theory to rural areas has not yet been made. We aim to address this gap in this chapter. In doing so, no detailed revision of the theory applied to the study area will be conducted but rather aspects of the spiral movement theory will be used to answer the following question: are new supply structures emerging in those regions where local food supply is thinning out? This question cannot be answered without analysing in greater detail the organizational structure of food retail and wholesale which are strongly vertically integrated in Germany.

Here, the weaknesses of older theoretical approaches become apparent: they strongly refer to the development of retail itself and the shopping behaviour of consumers but rarely take into consideration the underlying supply chains. There are several studies that have, since the 1990s, emphasized the importance of supply chains for the retail trade (see e.g. Coe and Hess, 2005). These studies use chain or network approaches such as the Global Value Chain framework (GVC, see for example Gereffi et al., 2005) or the theory of Global Production Networks (GPN, see for example Henderson et al., 2002). Even though these approaches are mainly used to analyse processes of globalization, we will use selected aspects of the GPN framework to build the theoretical framework of this chapter. On the one hand, we will use the network perspective when looking at the connection of retail development and the underlying supply chain relations. On the other hand we will use the three analytical categories *value, power,* and *embeddedness* (Henderson et al., 2002). Embeddedness is an important factor for a retail company's success as local consumption patterns and cultural norms have to be taken into account. Additionally, retail companies have to be locally embedded into supply chains and logistical infrastructure (Tacconelli and Wrigley, 2009).

The Development of Local Supply in Germany's Rural Areas

Retail trade in Germany, as in other European countries, has been undergoing processes of consolidation and concentration since the 1970s. These processes increase the dependence of consumers on the remaining companies. In Germany the top five German food trade companies accounted for 60 per cent of food sales in 2008 (Metro Group, 2009: 50). Such a high degree of concentration on the supply side of the trade means these companies have a strong position of power (Kuhlicke et al., 2005). At the same time, these processes of consolidation and concentration mean that the sales areas of individual stores increase while the number of shops declines and small retailers are successively cut out from the market. This affects especially those areas where the population's buying power falls below a certain level, as well as sparsely populated areas where there is not enough of a population in a supermarket's catchment area (Kuhlicke et al., 2005).

The clear winners of these concentration processes in Germany are retail chain stores. Integrated retail companies are part of larger companies or groups of companies that have their own wholesale operations. In this segment, so called *Verbundgruppen* are of particular importance. They are groups of cooperatives of mainly independent supermarkets. They generate 31.5 per cent of all retail sales (Wortmann, 2003: 4). In German food retail, Edeka and Rewe are the most important Verbundgruppen. Both are cooperatively organized enterprises (Wortmann, 2003).

However today, customers are often unable to tell the difference between these cooperatively organized retail companies, Edeka and Rewe, and regular retail chain stores as their structure and corporate strategy has changed significantly over the past years. Cooperatively organized retail companies have worked towards a homogenization of their membership structure and in this process, more and more independent retailers were ousted from these companies, most of them owners of small, traditional retail shops. They were replaced by larger chain stores that are partly run by a branch manager as an employee of the mother company. Often these centrally controlled stores are given to independent retailers again later on. However, one of these large chain stores can replace a number of small supermarkets. This trend is accompanied by an increasing integration of wholesale and retail trade which is also referred to as vertical integration. Wholesale trade is creating the connection between producers and consumers and producers and retailers. Wholesalers do not only trade products, they also store and transport them. Today, these typical tasks of wholesale trade are becoming increasingly integrated into the business of retail companies. In Germany, this process of integration was already much advanced in the 1970s (Wortmann, 2003).

Besides the purchase of merchandise, Edeka, Rewe, and other groups of companies with similar concepts have over time assumed more and more functions, such as the handling of payments and accountabilities towards suppliers. Later, more service-oriented functions such as consulting, finance and product development services for retailers were added (Wortmann, 2003). For small retailers in rural areas these service-oriented functions are very important, e.g. for obtaining financial support to renovate their stores, and for getting new ideas of how to present their merchandise or adjust their product range. Being a member of a Verbundgruppe also means that merchandise can be supplied at lower prices, an advantage that can be passed on to the consumers. Retailers that are not integrated into such a compound structure are struggling (Wortmann, 2003).

According to Kuhlicke et al. (2005), small, full-range traders such as small supermarkets and self-service stores are most important for local supply, besides the traditional bakers and butchers that have outlets in a large number of locations.

Comparing numbers of food retail stores in Germany from 2006 and 2011, the share of large supermarkets (400–999 m²) in the total number of food retail chain stores shrank from 14.8 per cent to 14.2 per cent, the share of small supermarkets (100–399 m²) shrank from 27.4 per cent to 18.7 per cent. At the same time the share of very large supermarkets and hypermarkets (1000 m² and more) increased from 16.3 per cent to 19.7 per cent and the share of discounters increased from 41.5 per cent to 47.4 per cent (Metro Group, 2012: 36).

Development of the Retail Sector in the Rural Areas of Central Hesse

The trend towards an increase in retail sales area and a reduction of store locations has also been confirmed by interviewees in our study area Central Hesse. The region is predominantly rural with an average population density of 194 inhabitants per square kilometre (Statistik Hessen, 2011: no page) and a lower than national and state average purchasing power (see Table 8.1).

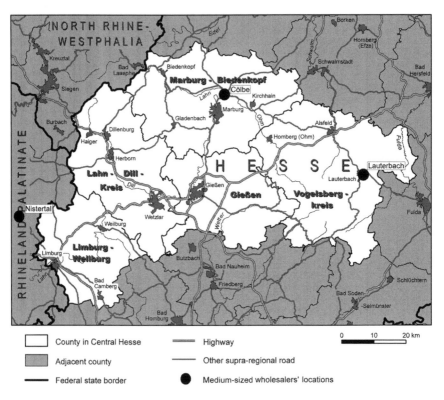

Figure 8.1   Central Hesse region and wholesale traders' locations
Source: Base map: Bundesamt für Kartographie und Geodäsie.

Table 8.1 Per capita purchasing power in Central Hesse's five counties

	Euros	**Purchasing power index (Germany = 100)**
Lahn-Dill	20,615	97.2
Gießen	20,486	96.5
Limburg-Weilburg	20,396	96.1
Marburg-Biedenkopf	19,618	92.5
Vogelsberg	19,542	92.1
Hesse	22,611	106.8

Source: GFK, 2013.

In Central Hesse, political decision makers at county and municipal level regard small-scale retail as crucial in maintaining local food and services supply and local social structures. A main challenge for these stores is their economic sustainability. A number of village shops have failed in the past because of the shopping behaviour of the local population who prefer to get their supplies at larger retail outlets outside the settlements with a broad product range and allegedly lower prices. Local suppliers suffer under this trend especially in those areas where a large number of people commute to work daily and combine their shopping with their trips to or from work. Thus, it can be said that the basic requirement for a functioning local supply is the willingness of the local population to buy at their local store. Older or less mobile people are more likely to do so because of different social needs and/or limited capacity to cover greater distances. But since only a relatively small part of the population is using local supply options because of such needs, more and more shops, especially in peripheral locations, close and more and more villages and small towns offer no shopping possibility at all. In some particularly affected areas this has led to the emergence of severe gaps in local supply.

This critical situation is worsened by the fact that 20 per cent of the interviewed shop owners of small supermarkets in Central Hesse are aged 60 years and above and 44 per cent of the shop keepers are planning to close their stores upon their own retirement. Only 18 per cent of shop keepers expect that a family member will continue their retail business and another 18 per cent are planning to sell their store. Taking these numbers into account and the reality of demographic change which decreases the absolute numbers of inhabitants and the purchasing power in many rural areas, there is a realistic threat that more and greater gaps in local food supply will emerge in the future.

However, the ageing of the local population might also lead to an increasing demand for local supply, based on the elderly's limited mobility. A similar effect on demand for local supply might come from higher fuel prices. If fuel prices are high, more consumers might chose to get their daily supplies locally, either because of economical or ecological motivations. Two out of three medium-sized wholesalers in Central Hesse stated that an increase in fuel prices is directly reflected in their turnover.

> When fuel prices rise [...] people do not drive so much. They rather think whether they should use their car to get some butter or milk. This can be positive for us, even though our logistics costs also increase. But it is definitely positive for our retailers. And if it is good for them, it is also good for us. (Interview with medium-sized wholesaler Giehl)

It can be said that 'smaller shops are emerging with distance as their main parameter in the competition' (Agergård et al., 1968: 35) is also true for Central Hesse. It can also be expected that demographic change with an increase of the elderly and less mobile population will have positive effects on local retailers' business. Indeed, interviewed wholesalers in Central Hesse stated that demographic change represents an opportunity rather than a threat to their business. Two of the three interviewed wholesalers mentioned that customers aged 60 years and above were central to their business and that recent increases in turnover could be primarily ascribed to this group of customers whereas the younger population would rather do their shopping in larger towns and stores.

That said, it can be stated that the transformations in retail that lead to ever greater sales areas and fewer store locations do not have to lead to the emergence of areas without food supply. The spiral movement theory describes a possible pre-stage of a Food Desert as a supply vacuum left behind by concentration processes and shop closures of large retail chains. These vacuums can then be filled by small, innovative retailers (Agergård et al., 1968). But who are these actors that create the much needed supply structures apart from strongly vertically integrated large retail chains such as Edeka and Rewe? Are new organizational structures emerging besides the large wholesale and retail companies that have the potential to secure local supply in rural areas?

Structures of Supplying Small-Scale Retail

In reaction to up-scaling of food retail in Central Hesse, a reorganization of the supply system can be observed that focuses particularly on small, independent retailers and helps them safeguard their future survival and find a market niche. The companies supplying these small retailers can be divided into three groups: (1) the large, cooperatively organized supplier companies (Edeka and Rewe); (2) medium-sized wholesalers (Gutkauf, Eberhardt, Giehl); (3) local and regional producers (the traditional craft-based food industry and farmers).

While the larger, cooperatively organized supplier companies such as Edeka and Rewe are in a process of restructuring towards strongly vertically integrated enterprises with a homogenous membership base, they successively get rid of their smallest retailers. Admittedly, these two groups of companies do also have concepts for smaller stores such as 'Rewe Nahkauf'. But these concepts are designed for stores with a minimum sales area of 300 m². In most cases, only stores that are larger than 600 m² are being integrated into Rewe's 'Nahkauf' concept (interview with Rewe Nahkauf). Another major retailer, Tegut, developed the 'Nah&Gut' concept which is being advertised with the label of local food supply. However, also in the case of this concept, stores have sales areas between 600 and 700 m² and most of them are even located outside the town centres (interviews with Nah&Gut shopkeepers).

The restructuring processes in these two groups of companies – Edeka and Rewe – creates the problem for both companies that they cannot easily exclude small retailers from their supply structures as they are embedded into the companies' cooperative structures. Therefore, from 2007 onwards, both companies have developed a cooperation model with medium-sized wholesalers that assume the supply of the smallest retailers.

> Rewe and Edeka supply retail stores that have a minimum sales area of 600, 700 m² and only because these have been associates in the group for many years. If, because of age reasons, the shopkeeper is about to change, Edeka might decide not to supply these stores any longer. Then Edeka approaches us and offers us to supply this retailer instead. We are in direct contact with Edeka. They tell us: "This shop will be off our list soon, go and see if you can supply him in the future." (Interview with medium-sized wholesaler Gutkauf)

The wholesalers that cooperate with Rewe and Edeka in that way are medium-sized, family-operated wholesalers. They traditionally supply small-scale retail in rural areas. The interviewed wholesalers supply between 120 and 450 stores between two and three times a week. All three wholesalers stated that the majority of their customers are small retail outlets with sales areas ranging between 100 and 300 m² and that they also supplied even smaller outlets. Which areas the wholesalers cover depends on their company location (see Figure 8.1). These locations have emerged historically. While Rewe and Edeka are active all over Hesse, each of the medium-sized, family-operated wholesalers covers only parts of the area of Central Hesse. Their market areas partly overlap inside Central Hesse and each of them also covers areas outside of the region. For example the wholesaler Giehl which is situated in Nistertal is active in Central Hesse, the northern parts of Rhineland Palatinate and the south of North Rhine-Westphalia.

The new model of cooperation to supply small-scale retailers in rural areas exists between Edeka and wholesaler Gutkauf as well as between Rewe and wholesaler Eberhardt. In both cases the concrete structure of cooperation is slightly different. Beyond that, both, Edeka and Rewe also cooperate with wholesaler Giehl who, in contrast to the other two wholesalers, organizes the procurement of goods autonomously. To do so, wholesaler Giehl is connected to the Markant Group. The Swiss company Markant is Europe's largest trade and services cooperation in food retail. As a classical buying syndicate, it is mainly active in the pooling of merchandise for their members (Markant, 2011).

In contrast to that, wholesalers Gutkauf and Eberhardt get their merchandise supplies directly from Edeka and Rewe. This network structure was labelled by one of the wholesalers as a 'piggy-back system'. It can best be explained by looking at the relationship between wholesaler Eberhardt and Rewe. Since 2009 the medium-sized wholesaler instead of Rewe has undertaken the supply of small retailers that have become too small to fit into the larger retail company's supply concept. However, the small retailers still order their merchandise with Rewe. Rewe then pools all merchandise going to small-scale retailers for the medium-sized wholesaler. The commissioning for individual retailers is then done by the medium-sized wholesaler. In the frame of this cooperation Rewe is simply outsourcing a segment of their logistics to the medium-sized partner while the company is still being represented at small retail outlets with its logo and name. While both Rewe as well as the medium-sized wholesaler are able to capture value in this relationship, retailers remain in direct contact with Rewe.

> Eberhardt's customers remain Rewe's customers just as any other customer but they have their own customer ID in order to differentiate them. They order just like any large Rewe supermarket through a mobile data collection device. All data are fed into Rewe's electronic data processing system and orders from these specific customer IDs are then being commissioned for one delivery to Eberhardt. Eberhardt then commissions the goods and supplies his customers the next day. These small customers also do not have a financial disadvantage, This system has been developed by us in the past two to three years. (Interview with Rewe Nahkauf)

An important prerequisite for such network structures to emerge and thus also for the positive developments in the areas of medium-sized wholesale is the cooperative structure of Rewe and Edeka. In both cases, shopkeepers of small retail stores – that have been associates of the Verbundgruppen for a long time – cannot be forced to give up their stores or be excluded from supply by Rewe or Edeka without greater damage to these companies' image or protest from small retailers. This historically-grown embeddedness forces the cooperatively organized companies to keep even small retailers in their supply structures. However, the cooperation with the medium-sized wholesaler allows them to keep efficient logistical structures.

> The shops we supply are all small "Nahkauf" – shops that do not reach a certain turnover level. For these stores, Rewe found a solution and does not have to state that they require a certain turnover minimum. This would not be very good for their image. It also has to be taken into account that Rewe is cooperatively organized and that especially amongst the small retailers there are still many associates. You cannot tell those shopkeepers: "You own your share in the

company but we cannot supply to you any longer." (Interview with medium-sized wholesaler Eberhardt)

Edeka also follows a similar model to the one between Rewe and medium-sized wholesaler Eberhardt. Edeka's partner is the medium-sized wholesaler Gutkauf. However, within the frame of this cooperation, this wholesaler is more independent than the wholesaler Eberhardt in its cooperation with Rewe. Gutkauf customers do not order their merchandise with Edeka but directly with the medium-sized wholesaler. The wholesaler is not integrated into Edeka's logistics system and buys a much lower share of merchandise from Edeka. While wholesaler Eberhardt purchases around 90 per cent of his product range directly from Rewe, the share is only 70–80 per cent in the case of wholesaler Gutkauf (interviews with both wholesalers).

> We are an independent wholesale company. We buy where we want and from whom we want. But we cooperate in such a way that we pick up a majority of goods from Edeka. Edeka commissions them over night and we pick them up the next day to commission it again for our customers to be delivered the same night or the next morning. (Interview with medium-sized wholesaler Gutkauf)

Through this model, larger retail companies such as Rewe or Edeka outsource logistics costs and optimize their business. In the case of Rewe, small retailers are at the same time retained as customers as is the name of the company as a local supplier because Rewe's logo remains visible in rural areas. For the medium-sized wholesalers this cooperation opens up contacts to potential new customers and the option to stabilize or even expand their business.

> Recently, Edeka's manager visited us. He emphasized very strongly that he highly appreciates our cooperation and that it is very important to Edeka to ensure local supply. To do so, they use partners such as us. (Interview with medium-sized wholesaler Gutkauf)

> Through this cooperation I have not only saved my business, I also got the chance to acquire a large number of new customers in 2009. These were retailers that came originally from Rewe and that I included into my service. (Interview with medium-sized wholesaler Eberhardt)

Also the small retailers integrated into this model profit because the medium-sized wholesalers do not only supply them with the merchandise needed, they also get access to a comprehensive range of services. Such services might encompass the analysis of market conditions, drafting of new store concepts, advertising and marketing, as well as advice regarding the product range. This is especially important taking into account that small retailers often do not have sufficient capacities and know-how to make necessary adaptations themselves or to design their stores in a more appealing or senior-pleasing way. Such ideas and concepts are being brought in and implemented though the medium-sized wholesaler. In return, the medium-sized wholesalers expect their customers to procure almost all products in their assortment through them, but allowing for additions with local or regional products.

Conclusion

Local food supply in rural areas has been undergoing structural and format changes for the past few decades. These changes have led to a situation where ever more and larger areas do not have shopping possibilities within easy reach anymore. The trend to scale up amongst the large integrated retail cooperatives like Edeka and Rewe is having a particularly strong impact. These companies are restructuring their businesses towards ever greater sales areas and increasingly lose interest in supplying small-scale retailers. After years of concentration also in the area of food wholesale, these small retailers face the threat of being excluded from competitive supply structures. However, the example of Central Hesse has shown that this supply gap can be filled by medium-sized wholesale companies. Similar to what Agergård et al. (1968) described with reference to urban areas, in the study area Central Hesse new organizational forms of local supply have emerged as a reaction to structural changes in the food retail sector. These new structures do not only allow the medium-sized wholesalers to stabilize or even expand their business, they also contribute to securing local supply in rural areas.

The positive future prospects of medium-sized wholesalers in the study region are also linked to a change in demand caused by demographic change and the higher attractiveness of local supply in times of rising fuel prices. The embeddedness of the medium-sized wholesalers into network relationships with Edeka and Rewe in the frame of the so-called 'piggy-back model' allows them to become part of these companies' logistics system and directly take over customers from them. This way companies that were earlier competing horizontally have now become vertically integrated. The medium-sized wholesalers now act as intermediaries and logistician between large retail cooperatives and small retailers. This model allows both Edeka and Rewe, as well as the medium-sized wholesalers, to capture value. For both retail cooperatives this model is a way to satisfy requirements that come with the historically-grown embeddedness of small retailers into the cooperative and avoid damage to their image. At the same time, the integration of the medium-sized wholesalers into their logistics system allows them to stay lean and efficient.

However, the strong network embeddedness of medium-sized wholesalers into the logistics network of Edeka and Rewe with its respective interdependencies and power relations also reveals the potential limits to the development of regional trade systems for small-scale retail. The question that arises is for how long larger partners such as Edeka and Rewe will regard the medium-sized wholesalers as a welcome supplement to their business and not as competitors. In addition to that, the survival of local supply also strongly depends on the shopping behaviour of the local population.

The case study has shown that, similar to Agergård et al.'s (1968) findings, spaces for new organizational forms of rural supply have emerged in the study area. However, the weaknesses of this theoretical approach have also become visible: it refers to developments in the retail sector and the shopping behaviour of consumers but it neglects the underlying supply chains. Therefore, it has only little or no explanatory value when looking at supply relations in the retail sector. However, in combination with chain or network approaches such as the Global Value Chain or the Global Production Networks approach, the spiral movement theory can still be used as an analytical framework and perhaps even build the foundation for the construction of new theories.

Acknowledgements

The data used in this chapter were collected by the authors within the frame of the project 'Integrated Preventive AAL Concept For the Aging Society in Europe's Rural Areas' (EMOTION-AAL) within the Ambient Assisted Living (AAL) Joint Programme of the European Union. This project was co-funded by the German Federal Ministry of Education and Research. Project partners were: ActiveSoft, BBraun Melsungen, the National Aeronautics and Space Research Centre of Germany (DLR), Diaconia University of Applied Sciences in Pieksämäki, the Retail Federation of Northern Hesse, the University of Applied Sciences Darmstadt, the Institute of Nanostructure Technologies and Analytics at Kassel University, Opsolution NanoPhotonics, Vitaphone, and the Department of Geography at Philipps-Universität Marburg. The German project management agency of the project was VDI/VDE.

References

Agergård, E., Olsen, P.A., and Allpass, J., 1968. *The Interaction between Retailing and the Urban Centre Structure – A Theory of Spiral Movement.* Lyngby.

Bartels, R., 1981. Criteria for Theory in Retailing. In: Stampfl, R.W. and Hirschman, E. (eds), *Theory in Retailing: Traditional and Nontraditional Sources.* Chicago, American Marketing Association.

Bloch, P.H., Ridgway, N.M., and Dawson, S.A., 1994. The Shopping Mall as Consumer Habitat. *Journal of Retailing* 70, 23–42.

Brown, S., 1987. An Integrated Approach to Retail Change: The Multi-Polarisation Model. *The Service Industries Journal* 7, 153–64.

Coe, N.M. and Hess, M., 2005. The Internationalization of Retailing: Implications for Supply Network Restructuring in East Asia and Eastern Europe. *Journal of Economic Geography* 5, 449–73.

Davidson, W., Bates, A.D., and Bass, S.J., 1967. The Retail Life Cycle. *Harvard Business Review* 54, 89–96.

Edeka, 2011. Die Geschichte einer Marke [online]. Hamburg, Edeka Zentrale AG and Co. KG. Available at: www.edeka.de/EDEKA/Content/Service/Impressum. jsp [accessed: 10 March 2011].

Furey, S., Strugnell, C., and Mcilveen, H., 2001. An Investigation of the Potential Existence of 'Food Deserts' in Rural and Urban Areas of Northern Ireland. *Agriculture and Human Values* 18, 447–57.

Gereffi, G., Humphrey, J., and Sturgeon, T., 2005. The Governance of Global Value Chains. *Review of International Political Economy* 12, 78–104.

GFK, 2013. GfK Kaufkraft Deutschland 2014 [online]. Bruchsal, GFK GeoMarketing GmbH. Available at: www.gfk.com/de/news-und-events/presse/ pressemitteilungen/Seiten/GfK-Kaufkraft-Deutschland-2014.aspx [accessed: 20 May 2014].

Gist, R.R., 1968. *Retailing Concepts and Decisions.* New York, John Wiley & Sons.

Globerman, S., 1978. Self-Service Gasoline Stations: A Case-Study of Competitive Innovation. *Journal of Retailing* 54, 75–86.

Heinritz, G., 1989. Der 'Wandel im Handel' als raumrelevanter Prozeß. In: Heinritz, G., Geipel, R., and Hartke, W. (eds), *Geographische Untersuchungen zum Strukturwandel im Einzelhandel* 63. Kalmünz.

——— 1991. Nutzungsabfolgen an Einzelhandelsstandorten in Geschäftsgebieten unterschiedlicher Wertigkeit. *Erdkunde* 45, 119–27.

Henderson, J., Dicken, P., Hess, M., et al., 2002. Global Production Networks and the Analysis of Economic Development. *Review of International Political Economy* 9, 436–64.

Hollander, S.C., 1966. Notes on Retail Accordion. *Journal of Retailing* 42, 29–54.

Izraeli, D., 1971. The Cyclical Evolution of Marketing Channels. *European Journal of Marketing* 5, 137–44.

Kirby, D.A., 1987. Village Shops: Improving their Chances of Survival. *Planning Practice and Research* 1, 16–20.

Kuhlicke, C., Petschow, U., and Zorn, H., 2005. Versorgung mit Waren des täglichen Bedarfs im ländlichen Raum. Studie für den Verbraucherzentrale Bundesverband e.V. Berlin, Institut für ökologische Wirtschaftsforschung, IÖW.

Kulke, E., 1992. Structural Change and Spatial Response in the Retail Sector in Germany. *Urban Studies* 29, 965–77.

——— 2006. Competition Between Formats and Locations in German Retailing. *Belgeo* 1–2, 27–39.

Markant, 2011. Herzlich willkommen bei der MARKANT AG [online]. Pfäffikon, Markant Handels- und Industriewaren-Vermittlungs AG. Available at: www. markant.com/dynasite.cfm?dssid=4656 [accessed: 20 March 2011].

Martenson, R., 1981. *Innovations in Multinational Retailing: IKEA in the Swedish, Swiss, German, and Austrian Furniture Markets.* Stockholm, Marketing Technology Center.

Mcnair, M.P. and May, E.G., 1976. *The Evolution of Retail Institutions in the United States.* Cambridge, Marketing Science Institute.

Metro Group, 2009. Metro-Handelslexikon 2009/2010. Düsseldorf, Metro AG.

———— 2012. Metro-Handelslexikon 2012/2013. Düsseldorf, Metro AG.

Miller, C.E., Reardon, J., and Mccorkle, D.E., 1999. The Effects of Competition on Retail Structure: An Examination of Intratype, Intertype, and Intercategory Competition. *Journal of Marketing* 63, 107–20.

Morganosky, M.A. and Cude, B.J., 2000. Large Format Retailing in the US: A Consumer Experience Perspective. *Journal of Retailing and Consumer Services* 7, 215–22.

Redondo, B.I., 1999. The Relation between the Characteristics of the Shopper and the Retail Format. *Marketing & Research Today* 28, 99–108.

Shaffer, H., 1973. How Retail Methods Reflect Social Change. *Canadian Business* 46, 10–15.

Statistik Hessen, 2011. Statistik Hessen [online]. Wiesbaden, Hessisches Statistisches Landesamt. Available at: www.statistik-hessen.de [accessed: 30 March 2011].

Tacconelli, W. and Wrigley, N., 2009. Organizational Challenges and Strategic Responses of Retail TNCs in Post-WTO-Entry China. *Economic Geography* 85 (1), 49–73.

Wortmann, M., 2003. Strukturwandel und Globalisierung des deutschen Einzelhandels. Available at: http://bibliothek.wzb.eu/pdf/2003/iii03-202a.pdf [accessed: 13 October 2013].

Chapter 9

Opportunities and Threats to the Development of Organic Agritourist Farms in Poland

Ewa Kacprzak and Barbara Maćkiewicz

Introduction

At the beginning of social and economic transformation in Poland, special opportunities for rural development were perceived as consisting, in particular, of the use of environmental assets and the existing agricultural structure of farming. The large number of private, small-scale, family-owned farms managed in a traditional way and the attractive and, to a significant degree, undeveloped environment are factors that should stimulate the development of agritourism (McMahon, 1996; Hall, 2004; Hegarty and Przezbórska, 2005; Drzewiecki, 2009; Sznajder et al., 2009). When promoting organic farming and the combination of such activities with tourism in organic agritourist farms, Jadwiga Łopata, who founded the European Centre for Ecological Agriculture and Tourism in Poland in 1993, claimed that 'Poland, in particular, has everything needed to follow an organic path of development. Beautiful landscape, natural resources, 2.5 million private farms and quite a lot of good farmers. Poland should become a democratic country with a good organic programme'. In this chapter an attempt has been made to evaluate the use of existing natural and cultural potential for the development of organic agritourism activities in Poland. Responses have been sought to the following question: is that development adequate to actual opportunities and needs, and if so, then to what degree? Apart from the quantitative particulars and the spatial distribution of organic agritourist farms, this chapter specifies the most essential conditions and factors that determine the development of organic agritourism activities. An attempt has also been made herein to assess the significance of organic agritourism activities in Poland. The source materials used herein are primary and secondary. The primary sources of information include the survey questionnaires that were sent to all farms that stated that they carry out organic agritourism activities in 2010. Prior to that, due to the absence of a single, up-to-date and complete list of such farms, a list of organic agritourist farms had been developed using the data made available by the Agricultural Consultancy Centre in Brwinów (Centrum Doradztwa Rolniczego), the European Centre for Ecological and Agricultural Tourism in

Poland and also on the websites of individual holdings. Among all 125 farms declaring themselves to be organic agritourist farms, 41 farms (33 per cent of the farms included in the list) completed the survey questionnaires and returned them. In addition, much primary information that has been significantly important for the research and its outcomes was collected during expert opinion surveys conducted with the representatives of governmental bodies and non-governmental organizations (NGOs) that coordinate the development of agritourism activities or organic agritourism activities in Poland – the Centrum Doradztwa Rolniczego (Agricultural Consultancy Centre) in Brwinów, the vice-president of the European Centre for Ecological and Agricultural Tourism (ECEAT) in Poland, the president of the Mazowieckie Wierzby Association of Agritourist and Organic Farms, and the organic food producers, as well as the owners of organic agritourist holdings. Apart from the abovementioned questionnaire surveys and expert opinion surveys, the most important research method applied in this chapter is SWOT analysis. This method allowed for the strengths and weaknesses of Polish organic agritourism activities to be determined and opportunities for their further development to be assessed. The secondary sources of information used in this chapter include, in particular, the Ministry of Agriculture and Rural Development, Chief Inspectorate for Trade Quality of Agricultural and Food Products (Główny Inspektorat Jakości Handlowej Artykułów Rolno-Spożywczych) in Warsaw, and the Agricultural Consultancy Centre (Centrum Doradztwa Rolniczego) in Brwinów.

Background

Rural Tourism, Agritourism, Organic Agritourism, and Ecotourism Activities in Poland

The broadest term covering phenomena related to the development of tourism in rural areas is rural tourism understood as leisure activities carried out in rural areas. It also encompasses various types of leisure activities related to nature, hiking, and sightseeing-orientated tourism, as well as cultural and ethnic tourism, which all take direct advantage of rural resources and values. The term 'agritourism activities', which is derived from Western European literature, was promoted in Polish academic publications and official documents in the 1990s (Drzewiecki, 2002; Kosmaczewska, 2007; Wojciechowska, 2009). Initially, agritourism activities were quite frequently identified with rural tourism activities, and both terms were used interchangeably. However, already in the mid-1990s more and more experts adopted the opinion that the term 'agritourism activities' covered a significantly narrower range than 'rural tourism activities' (Kacprzak, 2012). Defined by Wojciechowska (2009) as a form of rural tourism activity which encompasses specific tourism-oriented types of activities carried out on an operating farm, organized by a farming family with the use of resources available on their own farm and points of interest for tourists in the neighbourhood, as well as involving cooperation with the local community, and aimed at tourists seeking peace and quiet who have an interest in rural life and customs, agritourism activities are directly connected with an operating farm. The meaning of organic agritourism activities is even more limited as this term refers only to groups of farms that are authorized to simultaneously produce organic food and offer tourist services (Kuo and Chiu, 2006; Privitera, 2010). This understanding of organic agritourism activities conforms to the criteria applied by the industry (Agricultural Consultancy Centre in Brwinów and ECEAT Poland), according to which so-called organic agritourism activities represent agritourism activities that are carried out at a farm which is, at the same time, a producer of certified organic food.

In order to keep the terminology clear, it is also worth mentioning the definition of ecotourism activities. According to the Quebec Declaration on Ecotourism, which was formulated in 2002, they are a form of sustainable tourism activities that distinguish themselves from the broad concept of tourism development due to their active contribution to the protection of natural and cultural heritage, their engagement of local communities in planning and development and, therefore, contribute to the development of such communities, as well as due to the fact that they are addressed to individual tourists and tourists travelling in small groups. Ecotourism or green tourism activities include environmentally responsible travel and sightseeing in relatively untouched natural areas in order to achieve satisfaction and benefit from natural resources (and also any cultural values, both past and present, that are related to them), which promote environmental protection, have only a minor negative impact on the environment and create opportunities for the active social and economic engagement of local communities (Zaręba, 2006). In summary, it should be emphasized that all the terms defined above are strictly interrelated in that the scope of their meaning overlaps to a certain degree. Nevertheless, it should be borne in mind that they are different and cannot be equated.

In Poland, both farmers and tourists were initially interested in agritourism activities. That was related to the process of the transformation to a market economy which began in 1989. The changes in farming which resulted from the market economy and the restructuring of agriculture contributed to a significant decrease in the profitability of agricultural produce. This was triggered especially by a significant increase in prices of production means for agriculture and a decrease in prices of agricultural produce (and their instability), coupled with the disorganization of agricultural produce markets. The financial situation of the majority of farms deteriorated and, as a consequence, farmers started looking for additional sources of income to maintain or improve their current living standards. Some farmers with appropriate residential resources decided to take up agritourism activities. The key reason for commencing such activities was the desire to generate additional income (Kosmaczewska, 2007; Łejmel, 2001). Such initiatives were very compatible with the concept of the multifunctional development of rural areas which had already been implemented during the first years of transformation. However, as it turned out, the hopes initially invested in the development of agritourism activities were too high, as they were supposed to significantly contribute to the social and economic stimulation of rural areas – particularly those subject to the risk of marginalization.

After a relatively short period of euphoria, agritourism activities were approached in a more rational way (Kacprzak, 2012). At the moment, agritourism activities are regarded as an essential, however, not always the most important, direction for the diversification of farms and rural areas. Research demonstrates that the actual significance of agritourism activities for the development of rural areas is very often overestimated. This refers, in particular, to areas situated far from regular tourist destinations, where agritourism activities are still of marginal importance (Baum, 2011; Bednarek-Szczepańska, 2011). According to a report by Agrotec Polska Sp. z o.o. and the Stanisław Leszczycki Institute of Geography and the Spatial Organisation of the Polish Academy of Sciences (2012) there were 7,852 agritourism accommodation facilities in Poland in 2011 (Figure 9.1). They constituted a minor proportion, i.e. only 0.5 per cent of all individual agricultural premises with an area of more than 1 ha of agricultural land.

Countries of the European Union have been keenly interested in agritourism for several dozen years. Typically, in EU states agritourism activities are offered by several per cent of agricultural farms (Béteille, 1996b; Sikora, 2012). In Sweden as many as 20 per cent of farms provide accommodation to tourists. However, in Austria and Great Britain the figure is 10 per cent and 7 per cent, respectively (Butler et al., 1998), in Belgium and Germany less than 5 per cent and in France less than 2 per cent (Béteille, 1996a; Moinet, 2006). In Spain, where in contrast to other countries, agritourism activities are not strongly related to farming and in the majority of cases, taking into account rural accommodation facilities only, the share of this type of lodging in the total number of hotel places is small, amounting to only 4.4 per cent (Cánoves et al., 2005). It should be noted that in all the above-mentioned countries, the percentage of farms offering accommodation to tourists is, however, significantly higher than in Poland.

In Poland, agritourism activities started being combined with organic food production several years after the country's transformation to a market economy. The first initiatives related to the development of organic and agritourism activities in Poland were launched in 1993 when a project related to the promotion of organic and agritourism activities was initiated under the title of Eco-Agro-Tourism EAT (Łopata, 1993). The project was developed by an international group of ecologists and its aim was to propagate organic agritourism in Poland, Czech Republic, and Hungary. It was implemented by the ECEAT (European Centre for Eco-Agro-Tourism), a foundation established in 1993 in Amsterdam and sponsored by the ECC and the Dutch government. In Poland, 16 farms participated in the project. All of them were members of EKOLAND (Association of Organic Food Producers) or were in the process of switching to organic production.

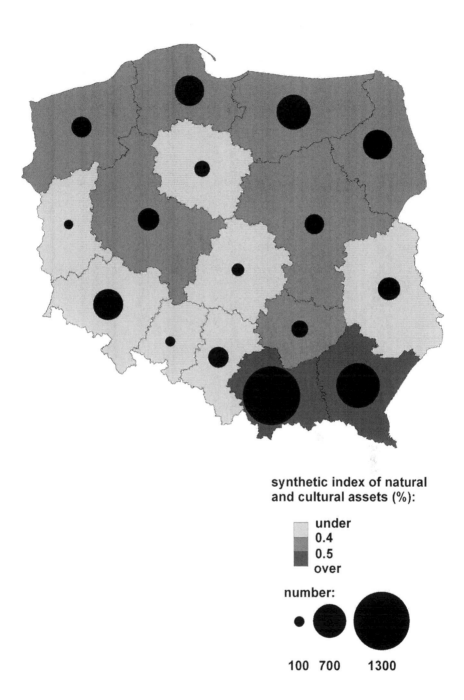

synthetic index of natural and cultural assets (%):

under
0.4
0.5
over

number:

100 700   1300

Figure 9.1   Number of agritourist farms in 2011

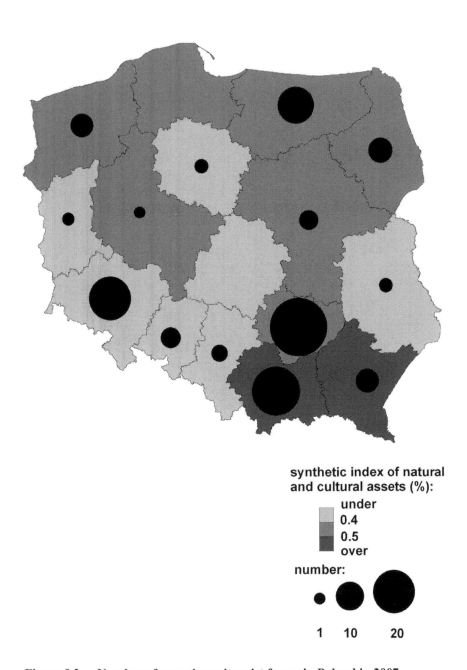

Figure 9.2   Number of organic agritourist farms in Poland in 2007

The ECEAT Poland was founded. This non-governmental organization promotes environmental protection, the cultivation of the traditions and culture of the Polish countryside, and in particular endorses organic agritourism, as well as education at farms, and has become the precursor of the development of organic agritourism activities in Poland. In 2002, which was declared the International Ecotourism Year by the United Nations Environment Programme (UNEP) and the World Tourism Organisation (UNWTO), the ECEAT Poland was awarded the Tourism for Tomorrow award sponsored by British Airways and the British Foreign Office for a project entitled 'Ecotourism on organic farms – Holiday on organic farms'. Despite those initiatives, the number of farms that carried out this type of activities was and remains very small. According to data from the Agricultural Consultancy Centre in Brwinów, there were 170 organic agritourist farms in 2007, which constituted less than 2 per cent of all agritourist farms (Figure 9.2). The research carried out demonstrates that currently this percentage is even lower, 1.6 per cent.

In the early twenty-first century, initiatives that followed Norwegian and Austrian models and consisted of the promotion of tourism in rural areas through ecotourism networks were introduced in Poland (Polak, 2013). In 2010, thanks to financial support granted by Iceland, Liechtenstein, and Norway under the Financial Mechanism of the European Economic Area and the Norwegian Financial Mechanism, as well as by the Republic of Poland under the Fund for Non-Governmental Organisations, the first and so far only model ecotourism network was established under the name 'Between the Bug and Narew Rivers' ('Między Bugiem a Narwią'). Under this project, the Social Eco Institute (Społeczny Instytut Ekologiczny) developed and patented the first eco-certification system for rural tourism facilities in Poland, i.e. the Polish Ecotourism Certificate, at the national level. This certificate is intended to implement and promote global ecotourism standards among those carrying out agritourism and rural tourism activities in Poland. The certificate is managed jointly by the Social Eco Institute and two agritourist associations, namely the Mazowieckie-Podlaskie Agritourist Association and the Mazowieckie Wierzby Association of Agritourism and Organic Farms. The certification system was developed for small and medium enterprises, above all those operating in rural areas and in areas with a high nature value. They may include farms and also any other accommodation facilities (maximum of 25 rooms), as well as hospitality facilities, shops, craftsmen, artists, guides, museums, cultural centres, sport centres and any other establishment providing services for tourists. In order to obtain the certificate, facilities and establishments must comply with over 100 strict ecology and ecotourism-related criteria (SIE, 2013). The criteria are extremely stringent, as a consequence of which, for example, only seven out of 28 facilities registered within the 'Between the Bug and Narew Rivers' ecotourism network have been granted the Polish Ecotourism Certificate. Simultaneously, five members of the network are organic agritourist farms and only two of them have been granted the Polish Ecotourism Certificate (Polak, 2013).

Organic Farming in Poland

According to the 2007 Action Plan for Organic Food and Farming in Poland for the years 2007–2013, the consumption of chemical means of production in Polish agriculture has always been lower than in the majority of European countries, and therefore the organic quality of agricultural production space in Poland and its biodiversity rank among the best in Europe. Such conditions are conducive to organic farming. A dynamic development of organic farming, delayed by two decades in comparison with the countries of Western Europe where it started growing in the 1980s, did not commence in Poland until the beginning of the twenty-first century (Brodzińska, 2010). At the beginning of social and economic transformations in Poland, organic farming developed relatively slowly, mainly due to insignificant domestic demand (Pawlewicz, 2007). In 1990, certificates for organic farms were introduced. At that time, there were only 27 organic farms operating on the market (Figure 9.3). The development of organic farming was very difficult with no state aid available for such activities. The potential to achieve higher selling prices was also limited because of poor organization of the organic produce market (Kłos, 2010). In the following decade, however, the number of organic farms slowly increased. In 1995, there were already 263 farms operating, and in 1999 their number doubled to 555 farms (IJHRS, 2001). Organic farming started to grow more rapidly in 1998 upon the introduction of subsidies compensating the costs of organic farm control and in 1999 upon the establishment of direct subsidies per 1 hectare of organic crops (Kłos, 2010).

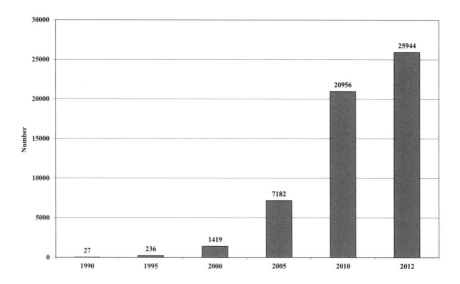

Figure 9.3   Number of organic farms in Poland between 1990 and 2012

The beneficiaries of these support mechanisms were farmers running organic farms or modifying their production profile from conventional to organic farming. A key event for the development of organic farming in Poland was the adoption of the Act on Organic Farming in 2001. As a consequence of the introduction of financial aid and appropriate legal regulations, the number of organic farms grew steadily. A year before the country's accession to the European Union, there were 2,286 organic farms operating in Poland (IJHRS, 2004). Nevertheless, a truly dynamic growth of interest in carrying out such activities was noted after the 2004 accession itself. From then onwards, the number of organic farms in Poland has been growing strongly and progressively. Simultaneously, rapid growth has been observed in the land area dedicated to organic farming. According to data from the Ministry of Agriculture and Rural Development, in 2012 there were 25,944 organic farms operating in Poland and the total area under organic cultivation was 661,687.30 hectares.

Many factors have contributed to this situation. The key factor, apart from the increasing demand for organic food or the expansion of potential markets, is certainly the fact that subsidies for organic farms are higher than subsides for conventional farming (Brodzińska, 2010; Kacprzak and Kołodziejczak, 2011; Pawlewicz, 2007). As the granting of subsidies to organic farming under the agricultural and environmental programmes within the Rural Development Plan is not conditional on the actual production of organic food for the market, it has to be assumed that one of the consequences of the implementation of such programmes is the existence of a certain number of 'fictitious' organic farms such as, for example, organic meadows and pastures, as well as grass on arable land (Pawlewicz, 2007). Simultaneously, it should be noted that despite the recorded increases in the number of organic farms over the last decade and the increase in the land area used for this type of activity, the percentage of agricultural land intended for organic farming in Poland in 2011 was 4.1 per cent, still lower than the average of 5.5 per cent estimated for the European Union (27 countries) (Eurostat, 2013). In terms of the share of the land area used for organic farming in the total area of agricultural land, Poland still differs significantly from European leaders such as Austria, Sweden, Estonia, the Czech Republic, and Latvia, where in 2011 the percentage was respectively 18.9 per cent, 15.7 per cent, 14.1 per cent, 13.1 per cent and 10.1 per cent. It is also worth noting that the proportion of organic farms in the total number of farms in Poland is not high. Although its value doubled in 2006–2011 and increased from 0.5 per cent to 1.0 per cent, the share of organic farms in the total number of farms remains insignificant (Ministry of Agriculture and Rural Development, 2012; Central Statistical Office, 2012). It should, however, be mentioned that the value of this share varies widely across the country. The proportion of organic farms in the total number of farms is the highest in the following voivodeships (NUTS 2 units): West Pomerania (6.5 per cent), Warmia-Mazuria (4.6 per cent), Lubuska Land (2.5 per cent), and Podlasie (2.3 per cent). On the other hand, organic farming is the least practised in the highly urbanized Silesia (0.2 per cent) and Łódź voivodeships (0.3 per cent), as well as in Opole (0.2 per cent) and Kujavia-Pomerania (0.4 per cent), known for their strong development of intensive farming (Figure 9.4).

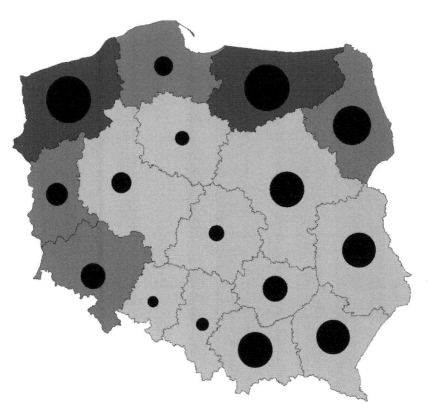

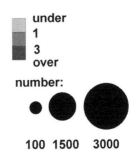

**organic farms in total number
of agricultural holdings (%):**

under
1
3
over

number:

100 1500   3000

Figure 9.4   Number of organic farms in 2011

The review demonstrates that very few organic farms also carry out agritourism activities. In 2011, out of the total of 23,449 farms of this type only 125 provided accommodation services to tourists (0.53 per cent). This means that a stay on an organic farm for tourism and leisure purposes is currently an uncommon form of tourism in Poland.

Organic Agritourist Farms in Poland: Description and SWOT Analysis

The group of farms that combine the production of certified organic foods with the provision of accommodation services to tourists in Poland is very small. To estimate the number precisely is quite difficult, as there is no single, up-to-date and complete register of such farms. A catalogue of organic agritourist farms can be found on the website run by the Agricultural Consultancy Centre in Brwinów that reports directly to the Minister of Agriculture and Rural Development. However, the catalogue has not been updated and the information it contains relates to 2007 (Centrum Doradztwa Rolniczego w Brwinowie, 2013). Another catalogue of organic agritourist farms is also available on ECEAT Poland's website. Although it is up-to-date, it only lists farms that are members of the association. It is interesting that a comparison of both catalogues demonstrates that only one third of farms who are members of ECEAT Poland have been included in the catalogue prepared by the Brwinów Agricultural Consultancy Centre. Furthermore, the Agroturystyka.pl website, which contains the largest selection of domestic agritourist accommodation facilities checked and recommended by the Gospodarstwa Gościnne Polish Federation of Rural Tourism, which is the most important organization specializing in the development of tourism in rural areas, has no functionality supporting an automatic search for organic agritourist farms among other farms. Organic agritourist farms registered with this website are all treated as agritourist farms, i.e. entities that do not engage in organic farming activities. Consequently, it is difficult to identify organic agritourist farms among the total of 1,636 farms listed. Therefore, the actual number of organic agritourist farms in Poland was verified on the basis of the combination of up-to-date lists of agritourist farms and farms that produce certified organic food products. It was thus determined that there were 125 organic agritourist farms operating in Poland in 2010, i.e. 45 farms fewer than specified in the 2007 list presented by the Agricultural Consultancy Centre (Nowak, 2010). It is worth noting, though, that despite the difference in the number of farms, their distribution was similar in both cases (figures 9.2 and 9.5).

The majority of organic agritourist farms were located in the south-eastern part of Poland, i.e. an area with the most attractive natural and cultural assets. A large number of such farms were also located in the equally attractive northeast (Warmia-Mazuria) as well as in Lower Silesia in the southwest. In total, more than half of all organic agritourist farms were located in those regions. On the other hand, the central part of Poland had very few or even no organic agritourist farms, as in the case of Łódź voivodeship. A comparison of the distribution of organic agritourist farms with that of organic farms demonstrated that some areas were characterized by both, a great number of organic farms and organic agritourist holdings. The correlation coefficient was significant, at 0.44.

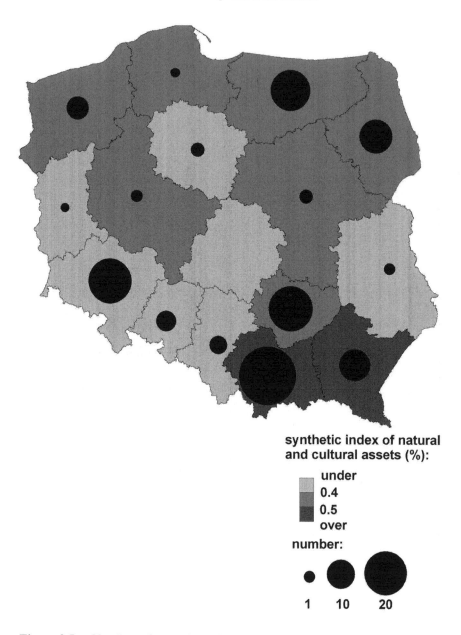

Figure 9.5   Number of organic agritourist farms in Poland in 2010

A comparison of the distribution of organic agritourist farms with that of agritourist ones showed them to be even more similar, the correlation coefficient amounting to 0.55. Although, as previously mentioned, the share of organic

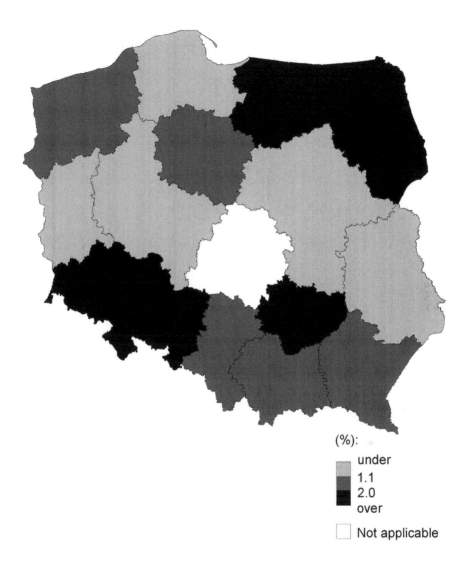

(%):

under
1.1
2.0
over

Not applicable

Figure 9.6   Share of organic agritourist farms in the total number of agritourist farms in 2010

agritourist farms in the total number of holdings carrying out agritourist activities was insignificant, at 1.6 per cent, in some voivodeships the proportion was significantly higher (Figure 9.6). This was observed in particular in Świętokrzyska Land and Opole, where the proportion of farms combining the production of organic food with the provision of accommodation services for tourists was 5.8 per cent and 5.5 per cent, respectively, of all the agritourist farms.

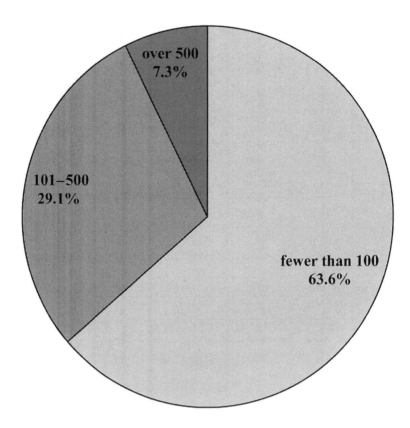

Figure 9.7   Area of organic agritourist farms

Questionnaires were sent out to a total of 125 organic agritourist farms, and replies were obtained from 41 farms (33 per cent). The overall area of the farms under study varied to a great extent. The largest farm, located in Warmia-Mazuria, occupied an area of 84.6 ha. The smallest, situated in Małopolska, had an area of 1.9 ha. The majority of farms (63.4 per cent) had an overall area not exceeding 15 ha (Figure 9.7); 17.1 per cent of farms occupied an area within the 15–30 ha range, while the proportion of the largest farms – covering more than 30 ha – amounted to nearly 20 per cent.

The vast majority of farms were established after 1990, i.e. after the transformation of the Polish political system. Only three farms out of the entire group under study had operated under the previous political regime, the oldest one dating back to 1964. The majority of respondents (65.3 per cent) belonged to the 41–65 age group. Also, a major proportion (nearly one-third) of respondents were young people between 18 and 40 years of age. There was only one person running an organic agritourist farm who was older than 65 years. The questionnaires provided evidence that organic farming/agritourism were occupations taken up by individuals with a decidedly higher level of education than an average resident in rural areas in Poland. A remarkably high proportion of respondents (30.6 per cent) had college or university education (Figure 9.8). The proportion was nearly twice as high as the national average (17 per cent) and three times higher than the proportion applicable to Polish rural areas which, in 2011, amounted to 9.9 per cent (Ludność. Stan i struktura demograficzno-społeczna. Narodowy Spis Powszechny Ludności i Mieszkań 2011, 2013). A similar trend in education level is also identified among Poles running agricultural farms. In 2010, only 10.3 per cent of all representatives of this group in Poland had college or university education (Pracujący w gospodarstwach rolnych. Powszechny Spis Rolny 2010, 2013). Furthermore, 36.5 per cent of respondents had completed secondary education and c. 33 per cent had graduated from vocational schools.

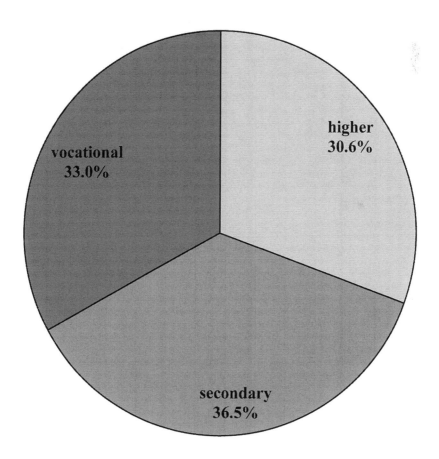

Figure 9.8   Education of operators of organic agritourist farms

The prevailing majority of respondents (75 per cent) declared they knew a foreign language (English and German being the most commonly mentioned). In many cases, operators of organic agritourist farms claimed they could communicate in several foreign languages.

Almost half (48.8 per cent) of all farms that filled in the questionnaire are members of professional associations related to their line of business. Interestingly, only five farms (12.2 per cent) belong to ECEAT Poland. Most respondents declared that their annual number of visitors did not exceed 100 (Figure 9.9). Questionnaire results also show that 29.1 per cent of farms host between 101 and 500 visitors a year. Only 7.3 per cent of farms regularly have more than 500 visitors during the year. The main visitor group comprises families with children, and the second largest group consists of pensioners. Adolescents usually visit organic agritourist farms in organized groups during field trips.

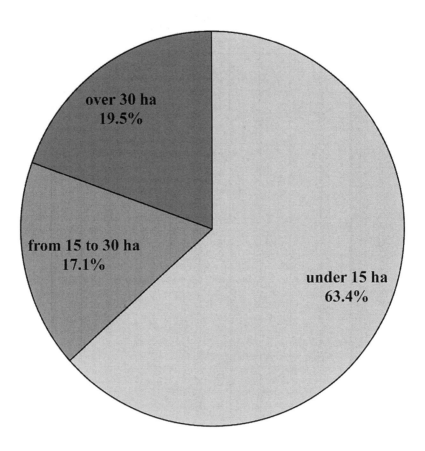

Figure 9.9   Number of visitors on organic agritourist farms over a year

It must be noted that none of the farms have ever provided services to people with disabilities, while barely 12 per cent of farms whose owners have replied to the questionnaire have facilities designed specifically for the disabled. The majority of farms (70.1 per cent) receive visitors from abroad. Foreign tourists come to Polish organic agritourist farms from a variety of countries including Germany (36.7 per cent, the largest group), the Netherlands (23.3 per cent), and Belgium (8.3 per cent). Austria, France, the UK, and Italy are also listed in the questionnaires as countries of origin of tourists, however the shares of all these nationalities in the overall group of foreign visitors do not exceed 5 per cent. As the survey shows, some of the farms have occasionally hosted visitors from Norway, Canada, USA, Spain, Sweden, Ukraine, Scotland, the Czech Republic, and Denmark. As for Poles using organic/agritourism services, the group is dominated by residents of large cities (Warsaw, Cracow, Łódź, Wrocław, Poznań, Katowice, and Gdańsk). Typical attractions offered to tourists in organic agritourist farms include carriage rides, horse-riding, and fishing. Some of the agritourism facilities, however, also provide an opportunity to participate in farming activities or in the process of production of eco-food (Wacher, 2007). An important fact to note is that the majority of respondents (63.4 per cent) declared they had their own unique '*specialty of the house*' products which they offered to visitors. The production of organic foods across the studied farms varied greatly. Results of the survey reveal that the main food items produced on their farms are cereals, fruit and fruit preserves, dairy products and vegetables. Half of all farms engage in meat production, usually as an activity complementing cattle-raising. Vegetable preserves and bread-baking are practised in one-third of all the farms participating in the survey, while meat products are made in one out of five establishments on average. A great majority of respondents (87.8 per cent) stated they offered visitors an opportunity to purchase food products from their organic agritourist farms. Managing an organic agritourist farm is the sole source of income for only 4.8 per cent of respondents. Most organic agritourist farm owners get income from other sources as well – for example by combining their organic agritourism business with the teaching profession, running another business or receiving pensions from the state. A review of forms of promotion demonstrates that most farms advertise on the internet. Over half of all the farms under study have their own website, those who do not advertise their businesses on other websites, e.g. those managed by their communes or districts. A frequently mentioned form of promotion is 'satisfied customers'. Many respondents said word-of-mouth marketing played an important role in their business. Interestingly, one of the farms listed entry in the guide, *Warmia and Masuria: Flavours and Aromas of Agritourism*, as an effective form of promotion.

Strengths

- natural and landscape values;
- presence of protected areas guaranteeing high quality natural environment;
- large forest areas;

- presence of areas with outstanding cultural resources;
- very high educational level of organic agritourist farm managers in comparison to the general rural population;
- widespread command of foreign languages among owners of organic agritourist farms;
- variety of tourism services provided by organic agritourist farms;
- variety of organic products and tourism attractions available at a single farm;
- rich and varied culinary offer;
- high number of farms with a so-called '*speciality of the house*';
- option to purchase organic products at the majority of farms;
- high quality of foods produced on the farms;
- high percentage of people working in agriculture and living in rural areas;
- large number of small and medium farms using traditional farming methods.

Weaknesses

- temperate climate;
- absence of up-to-date, available information about organic agritourist farms;
- lack of a strong national rural tourism organization supporting the development of organic agritourist businesses that would cooperate closely with the Ministry of Economy's Department of Tourism, Polish Tourist Organization, and national tourist authorities operating abroad;
- indolent attitude of local officials – e.g. agricultural advisory centres on the one hand endorse the development of rural tourism (organic farming/ agritourism included) and on the other tend to publish a lot of unreliable and unverified information about organic agritourist farms;
- lack of cooperation, e.g. in advertising, or membership in a joint organization of people running organic agritourist farms;
- lack of nationwide support for individuals who engage in organic agritourism;
- lack of government aid targeting organic agritourist farms (e.g. in the form of dedicated aid programmes);
- insufficient density of farms in the territory of the whole country, therefore, the farms are not accessible from all the large urban centres;
- significantly higher prices of products with an organic food certificate;
- almost complete absence of lodging facilities meeting the needs of people with disabilities.

Opportunities

- development of tourism, agritourism, and ecotourism;
- accession to the EU (EU funds and subsidies);
- increase in awareness of society in the area of organic and healthy food;
- eco fashion;
- good reputation of Polish food;

- consumer interest in the process of production and in the purchase of organic foods;
- interest of foreign tourists in Polish organic agritourist farms;
- significant workforce resources;
- traditional methods of farming;
- resilient operation of the Gospodarstwa Gościnne Polish Federation of Rural Tourism. With appropriate support (including financial aid) the Federation can become a robust organization efficiently supporting the development of organic/agritourism services in Poland.

Threats

- complicated and protracted procedure for obtaining an organic food producer certificate;
- obligation to renew the certificate on a yearly basis;
- no obligation to classify the quality of agritourism and organic and agritourism services;
- strong competition from agritourism facilities which often represent a form of rural tourism in real terms;
- excessive size of accommodation facilities in rural areas;
- absence of a high quality, unified training system for people providing tourism services in rural areas;
- excessive expectations of tourists: visitors are not willing to pay much but they expect a high standard of accommodation facilities, top quality organic foods, and an extensive range of recreational activities;
- low level of affluence of Polish society;
- deterioration in the financial condition of society and, as a consequence, decrease in the potential for buying more expensive organic products and spending holidays on organic agritourist farms;
- generally a low educational level of the rural population;
- tourists' limited knowledge of the countryside and of the typical operation of a farm, which makes some tourists dissatisfied with their stay in organic agritourist farms;
- poor knowledge and limited use of opportunities offered by the internet for managing organic agritourist farms;
- lack of understanding of tourists' needs, failure to match the needs and expectations of consumers (especially in terms of interior design and visual qualities).

Conclusions

Organic agritourism is still at a very low level of development in Poland despite the fact that there are favourable conditions for the functioning of organic agritourist farms – including natural and cultural values, dynamic growth in the number of organic farms and surface area of organic crops, as well as a growing interest in organically produced food.

There are still a number of significant limitations due to which organic agritourism development in Poland seems to be incommensurate with the existing natural and cultural potential. Perhaps the most important barrier to the growth in the number of such farms is the low level of affluence of the Polish people, which makes it impossible for many families to enjoy the tourist offer, nor does it allow them to purchase the more expensive certified organic food products. A further limitation which hinders the development of organic agritourism in Poland, and in fact all tourism, is the country's specific temperate climate, which practically restricts the use of organic agritourist farms to summer months only. What seems interesting, though, is the fact that, after Poland joined the EU, the Common Agricultural Policy has begun to play a vital role in its agriculture by supporting multifunctional development of rural areas and promoting organic farming, still this has not contributed to a greater increase in the number of organic agritourist farms. This is probably due to the fact that there have been no special government-initiated programmes targeted at this type of holding. Furthermore, there is no strong national rural tourism organization supporting the development of organic/agritourist businesses that would cooperate closely with the Ministry of Economy's Department of Tourism, Polish Tourist Organization, and national tourist authorities operating abroad.

There are significant spatial differences in the location of organic agritourist farms. Most of them are located in south-eastern Poland, i.e. in areas of the most outstanding natural and cultural richness. Many organic agritourist farms are also situated in the northeast, another region abounding in natural and cultural attractions, and in Lower Silesia. The central section of Poland has very few such farms. Also, what these farms have to offer, the quality of their services as well as prices differ widely, hence it is difficult to evaluate all those factors comprehensively, especially as there is no complete database of Polish organic agritourist farms. Among those whose operators returned the questionnaires, the majority have an area not exceeding 15 ha. Most of the farms receive fewer than 100 visitors during an average year. The majority declared they received guests from abroad. Interestingly, less than half of all holdings belong to associations related to the profile of their activity, with only a small proportion belonging to ECEAT Polska. The organic agritourist farm was the main source of income for a very small fraction of respondents. Other owners of organic agritourist farms combined this line of business with having a 'regular' profession (e.g. as a teacher) or running another business.

People involved in running organic agritourist farms are perceived as extremely resourceful individuals who are often referred to as the so-called 'rural elite'. They are characterized by a high level of education. This is closely related to the fact that some people have chosen to run such farms opting for a lifestyle change. By moving from a city to the countryside, educated city residents of outstanding ecological awareness have engaged in the business of organic agritourism.

Organic agritourism which meets the needs of city dwellers seeking a holiday close to nature, which provides knowledge of the rural lifestyle and the cultural potential of rural areas and which creates opportunities for profitable sales of farm-produced organic food, is certainly an additional source of income to these farms. However, to increase interest in this form of activity it is necessary, among other things, to provide financial support in the form of specialized programmes for organic agritourist farms as well as to set up a strong organization to promote the idea of combining the production of organically grown food with tourist services. In addition, organic agritourism will generate an attractive income once it starts offering a professionally developed tourist offer and engages in integrated promotional activity. It is possible that organic tourist farms will become the pride of Poland, but this will not be possible without implementing a well-drawn-up development policy for this type of tourist services.

Acknowledgements

We would like to thank all those who, by providing us with information, contributed to the creation of this chapter. In particular, we would like to express our thanks to Mr Sebastian Wieczorek, who is Vice-President of the European Centre for Ecological and Agricultural Tourism (ECEAT) in Poland, and to the employees of the Agricultural Consultancy Centre in Brwinów, as well as to Ms Barbara Polak, who is President of the Mazowieckie Wierzby Association of Agritourism and Organic Farms. We also owe our special thanks to our graduate Ms Joanna Karpińska and the farmers surveyed who devoted their precious time to completing surveys and giving interviews.

References

Agrotec Polska Sp. z o.o., 2012. Instytut Geografii i Przestrzennego Zagospodarowania im. Stanisława Leszczyckiego PAN. *Turystyka wiejska, w tym agroturystyka, jako element zrównoważonego i wielofunkcyjnego rozwoju obszarów wiejskich. Raport końcowy.* Available at: http://bip.minrol.gov.pl/ DesktopModules/Announcement/ViewAnnouncement.aspx?ModuleID=156 4&TabOrgID=1683&LangId=0&AnnouncementId=14984&ModulePosition Id=2199 [accessed: 2 May 2013].

Baum, S., 2011. The Tourist Potential of Rural Areas in Poland. *Eastern European Countryside* 17 (1), 107–35.

Bednarek-Szczepańska, M., 2011. Mit o agroturystyce jako szansie rozwojowej dla polskiej wsi. *Czasopismo Geograficzne* 82 (3), 249–70.

Béteille, R., 1996a. L'agritourisme dans les espaces ruraux européens. *Annales de géographie* 105 (592), 584–602.

———— 1996b. *Le tourisme vert.* Paris, PUF.

Brodzińska, K., 2010. Rozwój rolnictwa ekologicznego w Polsce na tle uwarunkowań przyrodniczych i systemu wsparcia finansowego. *Zeszyty Naukowe SGGW w Warszawie – Problemy Rolnictwa Światowego* 10 (25), 12–21.

Butler, R., Hall, C.M., and Jenkins, J., 1998. *Tourism and Recreation in Rural Areas.* London, Wiley.

Cánoves, G., Herrera, L., and Blanco, A., 2005. Turismo rural en España: un análisis de la evolución en el contexto europeo. *Cuadernos de Geografía de la Universidad de Valencia* 85 (77), 41–58.

Centrum Doradztwa Rolniczego w Brwinowie, 2007. *Gospodarstwa eko-agroturystyczne.* Available at: http://eko.radom.com.pl/bga/ [accessed: 10 May 2013].

Chrapek, G., 2007. Wpływ warunków przyrodniczych i pozaprzyrodniczych na rozwój ekoturystyki w regionie Polski południowo-wschodniej. In: Żabka, M. and Kowalski, R. (eds), *Przyroda a turystyka we wschodniej Polsce.* Siedlce, Wydawnictwo Akademii Podlaskiej, pp. 57–67.

Drzewiecki, M., 2002. *Podstawy agroturystyki.* Bydgoszcz, Oficyna Wydawnicza Ośrodka Postępu Organizacyjnego Sp. z o.o.

——— 2009. *Agroturystyka współczesna w Polsce.* Gdańsk, Wyższa Szkoła Turystyki i Hotelarstwa.

Hall, D., 2004. Rural Tourism Development in Southeastern Europe: Transition and the Search for Sustainability. *International Journal of Tourism Research* 6 (3), 165–76.

Hegarty, C. and Przezbórska, L., 2005. Rural and Agri-Tourism as a Tool for Reorganising Rural Areas in Old and New Member States – A Comparison Study of Ireland and Poland. *International Journal of Tourism Research* 7 (2), 63–77.

ECEAT Poland, 2013. Web page. Available at: http://www.eceatpoland.pl [accessed: 15 May 2013].

Eurostat, 2013. Web page. Available at: http://epp.eurostat.ec.europa.eu [accessed: 9 July 2013].

Główny Inspektorat Inspekcji Skupu i Przetwórstwa Artykułów Rolnych, 2001. *Rolnictwo ekologiczne w Polsce w latach 1999–2000.* Available at: http://www.ijhar-s.gov.pl/raporty-i-analizy.html [accessed: 20 July 2013].

IJHARS, 2004. *Rolnictwo ekologiczne w Polsce w 2003 roku.* Available at: http://www.ijhar-s.gov.pl/raporty-i-analizy.html [accessed: 20 July 2013].

——— 2007. *Raport o stanie rolnictwa ekologicznego w Polsce w latach 2005–2006.* Available at: http://www.ijhar-s.gov.pl/raporty-i-analizy.html [accessed: 20 July 2013].

Kacprzak, E., 2012. Rozwój agroturystyki w dolinie rzeki Warty. In: Młynarczyk, Z., Rosik, W., and Zajadacz, A. (eds), *Dolina rzeki Warty – przyrodnicze i turystyczne fascynacje.* Poznań, Wydawnictwo Naukowe Bogucki, pp. 103–26.

Kacprzak, E. and Kołodziejczak, A., 2011. Rozwój rolnictwa ekologicznego w Polsce w latach 2006–2009. *Biuletyn Instytutu Geografii Społeczno-*

ekonomicznej i Gospodarki Przestrzennej Uniwersytetu im. Adama Mickiewicza, Seria Rozwój Regionalny i Polityka Regionalna (14), 117–35.

Kłos, L., 2010. Rozwój rolnictwa ekologicznego w Polsce po wstąpieniu do Unii Europejskiej. In: Kryk, B. and Malicki, M. (eds), *Rolnictwo w kontekście zrównoważonego rozwoju obszarów wiejskich.* Szczecin, Wydawnictwo Economicus, pp. 48–66.

Kosmaczewska, J., 2007. *Wpływ agroturystyki na rozwój ekonomiczno-społeczny gminy. WWSTiZ w Poznaniu.* Poznań, Bogucki Wydawnictwo Naukowe.

Kuo, N. and Chiu, Y., 2006. The Assessment of Agritourism Policy Based on SEA Combination with HIA. *Land Use Policy* (23), 560–70.

Ludność, 2013. Stan i struktura demograficzno-społeczna. Narodowy Spis Powszechny Ludności i Mieszkań 2011. Warszawa, GUS.

Łejmel, K., 2001. Rola agroturystyki w rozwoju gospodarstw rolniczych i społeczności lokalnych. *Roczniki Naukowe SERiA* 3 (6), 69–73.

Łopata, J., 1993. Eko-turystka inaczej ... *Zielone brygady pismo ekologów* 50 (8), 39.

McMahon, F., 1996. Rural and Agri-Tourism in Central and Eastern Europe. In: Richards, G. (ed.), *Tourism in Central and Eastern Europe. 'Educating for Quality'.* Tilburg, University Press, pp. 175–82.

Ministerstwo Rolnictwa i Rozwoju Wsi, 2007. Plan Działań dla Żywności Ekologicznej i Rolnictwa w Polsce na lata 2007–2013. In: *Biuletyn Informacyjny Ministerstwa Rolnictwa i Rozwoju Wsi.* Warszawa, pp. 1–11.

Moinet, F., 2006. *Le tourisme rural.* Paris, France Agricole.

Nowak, J., 2010. Produkcja żywności ekologicznej w gospodarstwach eko-agroturystycznych. Maszynopis pracy magisterskiej.

Pawlewicz, A., 2007. Rolnictwo ekologiczne w Polsce – wybrane wskaźniki. *Zeszyty Naukowe SGGW w Warszawie – Problemy Rolnictwa Światowego* 17 (2), 415–22.

Polak, B., 2013. Prezes Stowarzyszenia Gospodarstw Agro i Ekoturystycznych Mazowieckie Wierzby. Personal interview, 24 May 2013.

Privitera, D., 2010. The Importance of Organic Agriculture in Tourism Rural. *Applied Studies in Agribusiness and Commerce* [online], 4 (1–2), 59–64. Available at: http://econpapers.repec.org/article/agsapstra/91113.htm [accessed: 9 May 2013].

Pracujący w gospodarstwach rolnych, 2013. Powszechny Spis Rolny 2010. Warszawa, GUS.

SIE, 2013. Web page. Available at: http://sie.org.pl [accessed: 5 May 2013].

Sikora, J., 2012. Agroturystyka. Przedsiębiorczość na obszarach wiejskich. Warszawa, Wydawnictwo C.H. Beck.

Sznajder, M., Przezbórska, L., and Scrimgeour, F., 2009. *Agritourism.* Wallinggford, CAB International.

Wacher, C., 2007. The Development of Agri-Tourism on Organic Farms in New EU Countries – Poland, Estonia, Slovenia. Report of a Winston Churchill

Travelling Fellowship 2006 [online]. Available at: http://orgprints.org/10860/1/ winstonchurchill.pdf [accessed: 5 June 2013].

Wieczorek, S., 2013. Wiceprezes polskiego oddziału European Centre for Eco Agro Tourism (ECEAT). Personal interview, 10 May 2013.

Wojciechowska, J., 2009. *Procesy i uwarunkowania rozwoju agroturystyki w Polsce.* Łódź, Wydawnictwo UŁ.

Zaręba, D., 2006. *Ekoturystyka.* Warszawa, Wydawnictwo Naukowe PWN.

Chapter 10

Agriculture in the NATURA 2000 Areas in Poland: Spatial Differences in the Absorption of Financial Means for Sustainable Development

Anna Kołodziejczak

Introduction

Since 2004, a new element of the conservation and management of natural resources in Poland has been the inclusion of some of its territory in the European Ecological Network NATURA 2000, which embraces special protection areas (SPAs) for birds and special areas of conservation (SACs) for habitats. In principle, economic activity can be carried on there without limitations, but on the condition that it does no harm to the habitats of the plants and animals for which the site has been established. Wherever possible, it is recommended to combine wildlife protection with human activity. While boasting particular landscape assets and great biodiversity, Poland runs the risk of its natural environment being damaged by road investments, industrial activity, and the intensification of agriculture. Because of the latter, mid-field tree clusters, small ponds and peatbogs tend to disappear from the landscape, together with the wild birds and other fauna those habitats support.

To encourage farmers to alter their land-use patterns, the European Union has introduced an agri-environmental programme. By conducting a specified type of agricultural activity in areas of particular natural value, farmers can participate in preserving the natural resources of their countries. When farming at NATURA 2000 sites, it is important to be aware of what can threaten them. The threats may result from a switch from the old land use to a new one or a change in the way the land has been used so far. The establishment of NATURA 2000 sites is a restrictive tool forcing the prevention or curbing of unfavourable modifications in the natural environment, i.e. keeping the right proportions in spatial management between its various economic, social and ecological functions.

The appearance of the NATURA 2000 network in rural areas has engendered interest among the academic circles as an expression of the European Union's interest in the protection of biodiversity. Studies have been made in many

European countries. The research carried out in Spain has focused primarily on the possibilities of financing and carrying out economic activity in NATURA 2000 areas (Rodriguez-Rodriguez, 2011). Similar studies have been performed in the eastern regions of Poland (Bołtromiuk and Kłodziński, 2011). In the opinion of many scholars, NATURA 2000 areas, in accordance with the network's assumptions, should offer a chance for ecological development rather than be a barrier to the sustainable development of agriculture and rural areas.

To emphasize how important the protection of biodiversity in agricultural areas is, economic instruments were employed under the agri-environmental programme for the years 2007–2013: subsidies granted to farmers differed according to whether their holdings were situated at NATURA 2000 sites or outside them.

The mechanisms of support for the coexistence of farming and nature protection in the NATURA 2000 areas as well as the rates offered in individual states differed widely. In the Czech Republic, the calculation of payment was based on the difference in income when the costs borne by farms included fertilization (80 kg/ha) and when they did not; the rate was 112 euros per ha of agricultural land. In Latvia support was only given for meadows and pastures, and the rate was 44 euros per ha of agricultural land. In Hungary the payment was less than 38 euros per ha of agricultural land because, again, financial support was only offered for meadows. In Poland, depending on the variant selected, the rates ranged from 141 to 356 euros per ha of agricultural land (Table 10.1).

Table 10.1   Rates of payment for packages 4 and 5* of the agri-environmental programme under RDP 2007–2013

Variants of packages 4 and 5	Rate of payment, euros per ha		Difference %
	Package 4	Package 5	
Protection of bird breeding habitats	307	351	14.2
Small sedge-moss communities	307	356	15.8
Tall sedge swamps	205	233	13.8
Litter meadows Molinion and Cnidion	307	356	15.8
Xerothermic grasslands	307	354	15.8
Semi-natural wet meadows	205	215	5.0
Semi-natural mesic meadows	205	215	5.0
Species-rich Nardion grasslands	205	223	6.7
Salt marshes	305	305	0
Natural lands	141	141	0

Note: * Package 4: 'Protection of endangered bird species and natural habitats outside of NATURA 2000 areas'; Package 5: 'Protection of endangered bird species and natural habitats in NATURA 2000 areas'.
Source: Bołtromiuk, 2011a: 368.

The location of an agricultural holding at a NATURA 2000 site can, therefore, entail some disadvantages, mostly in the form of loss of a part of income because of the limitations imposed on farming technologies and practices, while advantages are only ensured by the participation in and payments under the agri-environmental programme. It is important for farmers to have access to a comprehensive consultancy service on how to obtain subsidies for farming in harmony with the location of a holding at a NATURA 2000 site. A good example of agricultural consultancy in areas valuable in natural terms is Great Britain, where a non-governmental organization called the Farming and Wildlife Advisory Group was established; it assists farmers in acquiring funds for environmentally-oriented activities and changes their attitude towards the implementation of this type of activities. It is an organization run and financed by farm operators themselves, which helps them build the image of a farmer as a significant link in nature protection (Burchett and Burchett, 2011).

In Poland, of special significance for the conservation or restoration of natural values in the NATURA 2000 network were two measures of axis 2 of the Rural Development Programme (RDP) 2007–2013: improvement of the environment and the countryside, viz. agri-environmental payments and NATURA 2000 payments, as well as those under the Water Framework Directive implemented after conservation plans for a part of NATURA 2000 sites had been approved. At a NATURA 2000 site, only those elements of nature are protected that are listed in annexes to the directives. A farmer may join the programme of his own free will, but if he does, he must obey the rules set for the use of land, for example he cannot roll and harrow his meadows during the breeding season of birds. He is also obliged to cut grass only when their nesting season has ended, not earlier. The delimitation of a NATURA 2000 site in itself does not oblige the farmer to follow any special rules; a condition is his signing an agreement which sets rules of adjusting his agricultural activity at this site to conservation requirements. The farmer receives financial support depending on the scope of conservation. Those are not subsidies, however, but the transfer of means for continued use of the land and adjustment to measures specified in the programme. The programme is an opportunity, especially for those farmers who do not put their land to full use because it is not profitable for them to do so, which mainly concerns meadows and pastures. In other words, the NATURA 2000 programme provides the possibility of reintroducing agricultural production into those areas where production has been abandoned for its unprofitability.

Under the Rural Development Programme 2007–2013, the agri-environmental programme was modified on 1 March 2008 to include package 5, 'Protection of endangered bird species and natural habitats in NATURA 2000 areas'. This does not mean that before then NATURA 2000 sites were not given support. Under the environmental programme of the RDP 2004–2006, environment-related measures introduced by farmers at NATURA 2000 sites were rewarded with a 20 per cent bonus. This addition to the basic payment rates for which the farmer had to apply was in force for all the packages implemented on agricultural land located within the network. The application then had to be submitted together with a certificate issued by a voivode (head of a voivodeship, a NUTS 2 administrative unit) or the director of a national park, confirming the agreement between the plan of agri-environmental activity and the conservation plan for the NATURA 2000 site.

All the packages of the agri-environmental programme under the RDP 2007–2013 could serve to implement the principles of conservation of NATURA 2000 sites, but of special significance were those involving the protection of endangered bird species and natural habitats to preserve the biodiversity of rural areas at and outside NATURA 2000 sites. Under those packages, support was granted for selected bird species (10) and natural habitats (17) occurring on permanent grassland and listed in Appendix 4 to the Ordinance of the Minister of Agriculture and Rural Development of 28 February 2008. To protect endangered bird species and conserve valuable plant communities in those areas, its requirements specify the reduction in fertilization, the number and dates of cutting and control of the intensity of grazing. The package 'Protection of endangered bird species and natural habitats in NATURA 2000 areas' could be chosen by farmers whose holdings, or parts thereof, lie in such an area. Before he submitted an application, the farmer had first to learn if his meadow or sedge-moss community was situated in an area protected by the NATURA 2000 network. Next he had to contract an authorized expert to prepare suitable wildlife documentation in the year preceding the year when the agri-environmental obligation was supposed to start. The farmer was reimbursed for the cost of preparing this type of documentation with the first agri-environmental payment.

Later, the bonus for implementing pro-environmental measures at NATURA 2000 sites shrank in comparison with the solution adopted by the RDP 2004–2006. First, it was subsumed under a single agri-environmental package (package 5); secondly, the extra payment was reduced from 20 per cent to a maximum of 16 per cent (Bołtromiuk, 2011a).

Methodology

The chief aim of the research conducted was to identify the absorption pattern of payments under the package 'Protection of endangered bird species and natural habitats in NATURA 2000 areas' of the agri-environmental programme as an instrument inhibiting the use of natural resources in Poland in the years 2009 and 2010. Determining the level of absorption of EU funds supplied answers to many questions concerning the effect of this process on:

- the amount of financial means obtained by farmers whose holdings lie in the NATURA 2000 areas;
- spatial differences in the level of absorption of those payments;
- the extent of farmers' interest in those payments; and
- the development of extensive agricultural production.

This is important in the light of studies conducted in the EU states that sought to establish the role of Common Agricultural Policy instruments in protected areas in terms of sustainable rural development (Katona-Kovacs and Dax, 2008; Bastian et al., 2010). It is also of special significance for the conservation, or restoration, of natural values in the NATURA 2000 network.

A quantitative analysis was carried out on the basis of data obtained from the Agency for Restructuring and Modernisation of Agriculture (AR&MA), which has a regularly updated base concerning payments under the Rural Development Programme at the scale of voivodeships and poviats. The following variables were included in the study: number of applications granted, area of farmland receiving NATURA 2000 payments, amounts paid, area of NATURA 2000 sites, number of agricultural holdings, and mean area of farmland kept in good agricultural condition, which embraces both, favourable environmental conditions and the hectarage of agricultural holdings.

Using Pearson's correlation coefficient, the strength of linear dependence between the values of the diagnostic variables was tested. A high correlation was found to hold at the $\alpha = 0.05$ significance level between the number of applications granted and the number of farms ($r = 0.72$), as well as between the area of farmland covered by NATURA 2000 payments and the mean area of farmland kept in good agricultural condition ($r = 0.81$). Those data allowed the creation of the following indices:

- proportion of farms receiving NATURA 2000 payments in the total number of agricultural holdings;
- amount of payment per application;
- proportion of farmland embraced by the package 'Protection of endangered bird species and natural habitats in NATURA 2000 areas' in the total area of farmland maintained in good agricultural condition;
- amount of payment per ha agricultural land embraced by the package.

They were necessary to determine the absorption level with the help of a synthetic index. In its construction it refers to actual magnitudes of variables rather than to their ranks (Runge, 2007). When the data matrix consists of variables differing in units, their standardization is conducted prior to calculations, with zero corresponding to the average level in the country. The next step is calculating the arithmetic mean from the normalized values which, arranged in increasing or decreasing order, are interpreted in terms of the synthetic index (Racine and Reymond, 1977). Its values were calculated according to the formula:

$$W_s = \frac{\sum_{j=1}^{n} y_{ij}}{n}$$

where:

W_s – the synthetic index;
j – 1, 2, ..., n;
n – the number of variables taken into account;
y_{ij} – the standardized value of the *j*-th variable for the *i*-th object.

The results were divided into five groups, or intervals, with the following end points:

I   – a very low level of absorption of financial means: under -0.499;
II   – a low level of absorption: from -0.500 to -0.001;
III   – an average level of absorption: from 0 to 0.500;
IV   – a high level of absorption: from 0.501 to 1.000;
V   – a very high level of absorption: over 1.001.

Discussion

The NATURA 2000 network is a functionally consistent system of protected areas across the European Union. It includes:

- special protection areas for bird species (SPAs) identified in accordance with the recommendations of the Birds Directive;
- special areas of conservation for natural habitats (SACs) identified in accordance with the recommendations of the Habitats Directive.

The idea of the NATURA 2000 network rests on the assumption that the protection of natural habitats and species in network areas will take place within the conditions of economic land use conforming to the rules of sustainable development, i.e. of ensuring prosperity while conserving biological resources and biological diversity. What is crucial is the effect, viz. maintaining the so-called favourable conservation status of the types of habitat and species for which those areas have been set up, or its restoration. In the case of natural habitats, favourable conservation status means that the natural range of the given type of habitat does not shrink and that it keeps its specific structure and function and protects the species that are typical of it. In the case of a species, favourable conservation status means that its population does not dwindle, so it can survive in a biocoenosis in a longer time perspective, its natural range does not shrink, and the area of its habitat is large enough (Tworek and Makomaska-Tuchlewicz, 2008). In SACs, agricultural production can still be carried on, and the only ventures that must be limited are those which may have a detrimental impact on an area's natural features. The limitations involve primarily:

- choice of the structure of agricultural production;
- land-use pattern, size of crop-rotation fields;

- prohibition to turn meadows and pastures into arable land;
- methods and dates of land and crop cultivation;
- level, dates and forms of fertilization applied;
- location of the most intensive forms of agricultural production (hop plantations, orchards, garden crops, greenhouses, plastic tunnels, mushroom farms, pigs, poultry, fur animals);
- choice of plant protection chemicals and methods of their application;
- level of leaching of nitrogen and phosphorus compounds from agricultural land;
- introduction of alien species, plant varieties and animal races; and
- introduction of genetically modified plants and animals (Ilnicki, 2004).

In SPAs for bird species, agriculture-related threats can be found in abundance. They include:

- an increase in the level of fertilization which reduces the food base for birds, makes it difficult or impossible for them to move, and causes them to abandon their broods prematurely;
- use of pesticides impairing the reproduction capacity of birds or killing them directly and destroying their food base;
- intensive economic practices, grazing responsible for unfavourable changes in habitats and losses in broods;
- fallowing of grasslands which causes them to be overgrown with tall plants, thus changing the habitat;
- mechanization leading to the destruction of broods or nestlings during the first cut;
- flood control and draining practices which accelerate the process of grassland intensification and its transformation into arable land;
- an increase in the proportion of winter cereals, which reduces the food base for grain-eaters provided by the stubble left for the winter period (Liro, 2003).

The network of NATURA 2000 sites is highly diversified as to the modes of protection employed, and its chief tools are conservation plans, procedures for making environmental impact assessments, and suitable funding. In Poland the establishment of its sites has been a matter of much discontent over the recent years, both among local authorities and some members of society, because it was taken to mean limitations on development in those areas.

In 2010, the NATURA 2000 sites in Poland (except marine ones) occupied a total of 8,354,225 ha, or 26.7 per cent of the country's area. The network included:

- 144 special protection areas for bird species (SPAs), identified in accordance with the recommendations of the Birds Directive and occupying 4,922,367 ha;

- 823 special areas of conservation for natural habitats (SACs), identified in accordance with the recommendations of the Habitats Directive and occupying 3,431,858 ha.

The distribution of the NATURA 2000 sites in Poland is not regular. Their highest proportion can be found in Podlasie (55.7 per cent of its total area), West Pomerania (48.6 per cent) and Subcarpathia (47.7 per cent), and the lowest, in Opole (4.4 per cent) and Łódź (5 per cent).

A substantial proportion of the NATURA 2000 sites are situated on agricultural land, and this produces economic effects in farming. In the opinion of Zegar (2010), the effects are direct and indirect. The former include loss of economic benefits because of constraints on the structure of agricultural production and the intensity of farming, investment outlays necessary in areas with environmental limitations, and costs of preparing applications, plans and assessments of the impact of an agricultural holding on the environment.

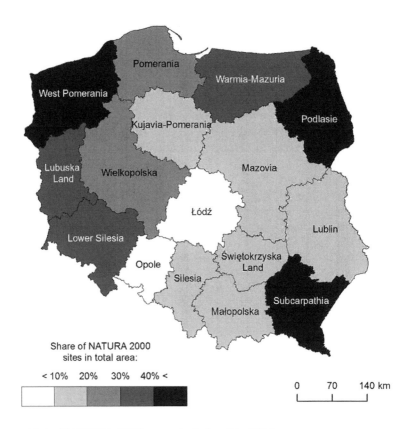

Figure 10.1   NATURA 2000 areas in Poland in 2010
Source: Environmental Protection, 2011. GUS.

This loss of advantages can be compensated for by payments designed for areas of agri-environmental limitations and by reduced inputs, including mineral fertilizers and plant protection chemicals. The latter include possible advantages from a balanced or ecological way of farming as well as those deriving from services based on environmental assets.

It was estimated at first that applications under this RDP 2007–2013 measure would come from 150,000 farms operating a joint area of 370,000 ha at NATURA 2000 sites (Rowiński, 2008). However, in 2010 applications were submitted by a mere 1,955 farmers, and the area of agricultural land covered by those subsidies was 44,808.91 ha, or 0.29 per cent of total agricultural land in Poland (Table 10.2).

As can be seen, farmers were not much interested in this form of financial assistance. This resulted directly from the complicated procedures involved and high starting costs, which was a barrier to small farms. The greatest numbers of applications were granted in the voivodeships of Podlasie, Lublin, Subcarpathia, and West Pomerania, and the smallest, in Opole, Silesia, Łódź, and Kujavia-Pomerania.

Table 10.2   Implementation of the package 'Protection of endangered bird species and natural habitats in NATURA 2000 areas' in Poland in 2010

Voivodeship	Applications granted	Agricultural land embraced by package		Amount	
		ha	%	In thous. euros	In euros per application
Lower Silesia	144	3,133.37	0.32	985.58	6,844.31
Kujavia-Pomerania	27	345.65	0.03	108.12	4,004.60
Lublin	253	4,587.98	0.32	1,649.70	6,520.54
Lubuska Land	175	6,398.24	1.42	2,238.25	12,789.98
Łódź	25	194.99	0.02	70.82	2,832.90
Małopolska	68	1,183.44	0.18	416.26	6,121.43
Mazovia	130	2,536.55	0.13	869.26	6,686.62
Opole	7	6.54	0.00	3.24	462.69
Subcarpathia	242	4,993.19	0.72	1,706.49	7,051.60
Podlasie	266	7,334.25	0.69	2,592.59	9,746.60
Pomerania	110	1,569.38	0.19	539.55	4,905.04
Silesia	16	506.01	0.11	181.82	11,363.82
Świętokrzyska Land	77	448.65	0.08	167.81	2,179.32
Warmia-Mazuria	111	2,687.08	0.25	955.65	8,609.52
Wielkopolska	130	2,385.72	0.13	834.39	6,418.38
West Pomerania	214	6,497.87	0.68	2,302.76	10,760.58
Poland	**1,995**	**44,808.91**	**0.29**	**15,622.31**	**7,830.73**

Source: Own calculations on the basis of data of the AR&MA Department of Information Systems Management.

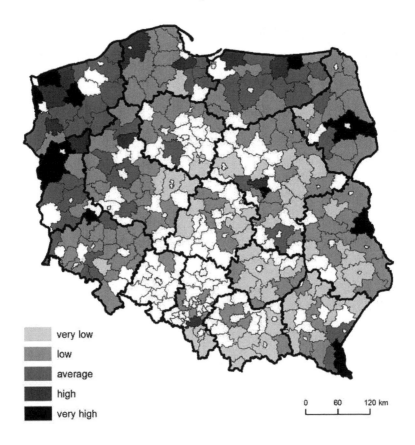

very low

low

average

high

very high

0   60   120 km

Figure 10.2   Absorption of NATURA 2000 payments in Poland in 2010
Source: Own compilation on the basis of AR&MA data.

The average area of agricultural land on a single farm for which payments were granted was 20.3 ha in 2010, as against 26.5 ha in 2009.

Absorption of those payments varied in space, sometimes falling short of a possible use of funds for maintaining agricultural production in NATURA 2000 areas.

A very high level of absorption of payments (group V) was recorded in 12 poviats (NUTS 4 units): Gorzów, Słubice and Sulęcin (Lubuska Land), Kamień Pomorski, Police and Łobez (West Pomerania), Głogów (Lower Silesia), Włodawa (Lublin), Pruszków (Mazovia), Ustrzyki Dolne (Subcarpathia), Białystok (Podlasie), and Węgorzewo (Warmia-Mazuria). The mean area of agricultural land maintained in good condition in those poviats ranged from 6.72 ha (Pruszków) to 37.14 ha (Łobez). This type of financial support on farms of average size (5–15 ha) was an encouragement for them to choose and continue an extensive farming system. On large farms located in the poviats of Lubuska

Land, West Pomerania and Warmia-Mazuria, those payments were a recompense for additional costs involved in the limitations imposed on farming and for the lost possibility of intensifying agricultural production.

Group IV with a high level of fund absorption included 11 poviats located mostly in western and northern Poland. As in group V, those were large and medium-sized farms maintaining agricultural land in good condition.

Group III with an average level of fund absorption was made up of 40 poviats located in Warmia-Mazuria, West Pomerania, Lubuska Land, Wielkopolska, Lower Silesia, and Subcarpathia.

The greatest number of poviats belonged to group II with a low level of absorption of financial means for agricultural production in NATURA 2000 areas. Here farms varied in size. Larger ones were concentrated in the poviats of Warmia-Mazuria, Pomerania, Kujavia-Pomerania, Wielkopolska, and Lower Silesia, while smaller ones extended along the eastern state border, i.e. in the voivodeships of Podlasie, Lublin, and Subcarpathia. A very low absorption level (group I) was recorded in 53 poviats, situated mostly in central and south-central Poland. Predominant in both groups were small farms where NATURA 2000 payments were an encouragement to continue extensive crop culture and animal husbandry, and to take environmentally oriented steps in the form of the desired agri-technical services.

In poviats with the most extensive NATURA 2000 areas in West Pomerania, Warmia-Mazuria, and Lubuska Land, farms belonged to the groups with a high and a very high level of absorption of the payments. In the poviats of Podlasie and Subcarpathia, also with the most extensive NATURA 2000 areas, farms fell into the groups with an average and a low level of absorption.

The above data for 2010 show there to be little interest in payments connected with the location of farms at NATURA 2000 sites. The procedures described earlier show that the many conditions that have to be met and the requirement for a farmer to have his own means to start an activity in order to obtain a payment clearly inhibit this process. Also, there are few experts authorized to prepare wildlife reports which prolongs the procedure of application submission.

Another reason is farmers' scant knowledge about the NATURA 2000 package. This was corroborated by a study carried out by the Polish Bird Protection Association in a NATURA 2000 area, the Ostoja Warmińska in Warmia-Mazuria voivodeship. Half of the respondents knew the advantages of running a farm in a NATURA 2000 area, 41 per cent did not, and 9 per cent did not respond. A mere 6 per cent of the respondents gave an affirmative answer to the question of whether NATURA 2000 helped them to conduct agricultural activity, while 22 per cent thought that NATURA 2000 hampered it. As many as 47 per cent had no opinion on this matter and 25 per cent thought that NATURA 2000 neither helped them nor hindered their agricultural activity.[1] This shows that, drawing on the experiences

---

1   'Farmers do not know the NATURA 2000 programme', *Wprostweekly*, 25 December 2008. Available at: http://www.wprost.pl/ar/148452/Rolnicy-nie-znaja-programu-Natura-2000/, accessed 4 June 2014.

of the West European states, agricultural consultancy centres in Poland should engage in meetings, training courses, and disseminating information on the internet to reach farmers whose holdings lie in the NATURA 2000 areas in order to boost their interest in those payments.

Conclusions

The NATURA 2000 network is not, by assumption, a system directed against man, but designed to control his activity in the environment. Even within the network of bird and habitat protection areas it is possible to conduct economic activity that is not destructive to nature. The agri-environmental programme implemented at the NATURA 2000 sites seeks to put in order the way farming is practised in those valuable areas. The differences occurring between the poviats in the level of package implementation follow from the possibilities offered by their various environmental conditions as well as the skills and level of awareness of their farmers. The environmental impact of the community's programmes and financial support depends not only on the location of an agricultural holding, but also on its size and on a laborious and costly preparation of documentation for the interested farmer, identifying a protected species or habitat on his land. As follows from direct contacts with the beneficiaries of the agri-environmental programme, they expect more complex support on the part of consultants and a simplification of the administrative procedures involved which are cumbersome not only for the beneficiaries but for the programme administrators as well. Even though the European Union has designed an efficient financial instrument inhibiting the use of natural resources in the form of its agri-environmental programme, natural values of the Polish countryside still suffer degradation. The programme is totally lacking in mechanisms protecting unproductive elements of the agricultural landscape, like weeds, mid-field ponds, mid-field and waterside tree clusters, or marshy land and meadows situated along ditches and rivers. This leads to loss of habitats for plants that provide a food base for animals. As a result, farmers liquidate those sanctuaries of biological diversity and put them to productive use, since they have no motivation to keep them – unproductive areas are excluded from direct payments and neither is there an agri-environmental payment for them. An adverse effect on the 2010 level of absorption of the payments was also produced by an unfavourable change in funding that lessened the economic stimulus. This is especially readily visible in central Poland.

References

Bastian, O., Neruda, M., Filipova, L., et al., 2010. Natura 2000 Sites as on Asset For Rural Development: The German–Czech Ore Mountains Green Network Project. *Journal of Landscape Ecology* 3 (2), 41–58.

Bołtromiuk, A., 2011a. Koncepcja systemu publicznego wsparcia rolnictwa na obszarach europejskiej sieci ekologicznej (Conception of A Public Support System for Agriculture in Areas of the European Ecological Network). In: Bołtromiuk, A. and Kłodziński, M. (eds), *NATURA 2000 jako czynnik zrównoważonego rozwoju obszarów wiejskich regionu Zielonych Płuc Polski*. Warszawa, Instytut Rozwoju Wsi i Rolnictwa PAN, 359–97.

——— 2011b. Przesłanki utworzenia i geneza sieci NATURA 2000 (Factors and Genesis of Establishing the NATURA 2000 Network). In: Bołtromiuk, A. and Kłodziński, M. (eds), *NATURA 2000 jako czynnik zrównoważonego rozwoju obszarów wiejskich regionu Zielonych Płuc Polski*. Warszawa, Instytut Rozwoju Wsi i Rolnictwa PAN, 71–103.

Bołtromiuk, A. and Kłodziński, M. (eds), 2011. *NATURA 2000 jako czynnik zrównoważonego rozwoju obszarów wiejskich regionu Zielonych Płuc Polski* (NATURA 2000 as a Factor of the Sustainable Development of Rural Areas in Poland's 'Green Lungs' Region). Warszawa, Instytut Rozwoju Wsi i Rolnictwa PAN.

Burchett, St. and Burchett, S., 2011. *Introduction to Wildlife Conservation in Farming*. John Wiley & Sons.

Council Directive 79/409/EEC of 2 April 1979 on the conservation of wild birds.

Council Directive 92/43/EEC of 21 May 1992 on the conservation of natural habitats and of wild fauna and flora.

Ilnicki, P., 2004. *Polskie rolnictwo a ochrona* środowiska (Polish Agriculture and Environmental Protection). Poznań, Wydawnictwo Akademii Rolniczej im. Augusta Cieszkowskiego.

Kaługa, I., 2009. *Korzyści dla rolnictwa wynikające z gospodarowania na obszarach NATURA 2000* (Advantages for Agriculture From Farming in NATURA 2000 Areas). Warszawa, Ministerstwo Środowiska.

Katona-Kovacs, J. and Dax, T., 2008. Sustainable Rural Development in Environmentally Protected Areas of Hungary and Austria: The Role of CAP Payments. 12th Congress of the European Association of Agricultural Economists – EAAE, pp. 1–5.

Liro, A., 2003. Sieć NATURA 2000 a zrównoważony rozwój obszarów wiejskich (The NATURA 2000 Network and Sustainable Development of Rural Areas). In: Makomaska-Juchniewicz, M. and Tworek, S. (eds), *Ekologiczna sieć NATURA 2000. Problem czy szansa*. Kraków, Instytut Ochrony Przyrody PAN.

Ochrona środowiska, 2011 (Environmental Protection, 2011). Warszawa, GUS.

Ordinance of the Minister of Agriculture and Rural Development of 28 February 2008 on detailed conditions and mode of granting financial assistance under the measure 'The Agri-Environmental Programme' under the Rural Development Programme for the years 2007–2013 (*Official Gazette* no. 34, position 200).

Ordinance of the Minister of Agriculture and Rural Development of 27 October 2008 altering the Ordinance on NATURA 2000 special protection areas for birds (*Official Gazette* no. 196, position 1226).

Ordinance of the Minister of Agriculture and Rural Development of 26 February 2009 detailed conditions and mode of granting financial assistance under the measure 'The Agri-Environmental Programme' under the Rural Development Programme for the years 2007–2013 (*Official Gazette* no. 33, position 262).

Program Rozwoju Obszarów Wiejskich 2007–2013 (Rural Development Programme 2007–2013), 2007. Warszawa, Ministerstwo Rolnictwa i Rozwoju Wsi.

Racine, J.B. and Reymond, H., 1977. *Analiza ilościowa w geografii* (Quantitative Analysis in Geography). PWN, Warszawa.

Rodriguez-Rodriguez, D., 2011. *Natura 2000 Network and Rural Development: Current Situation and Future Perspectives*. Lambert Academic Publishing.

Rowiński, J., 2008. Program Rozwoju Obszarów Wiejskich na lata 2007–2013 (Analiza zatwierdzonej wersji programu i pierwszych lat realizacji) (Rural Development Programme for the Years 2007–2013: Analysis of the Approved Version of the Programme and the First Years of Its Implementation). *Ekonomiczne i społeczne uwarunkowania rozwoju gospodarki żywnościowej po wstąpieniu Polski do Unii Europejskiej* nr 118. Warszawa, Instytut Ekonomiki Rolnictwa i Gospodarki Żywnościowej-PIB.

Runge, J., 2007. *Metody badań w geografii społeczno-ekonomicznej – elementy metodologii, wybrane narzędzia badawcze* (Research Methods in Socio-Economic Geography: Elements of Methodology and Selected Research Tools). Wydawnictwo Uniwersytetu Śląskiego, Katowice.

Tworek, S. and Makomaska-Tuchlewicz, M., 2008. Podstawy prawne sieci NATURA 2000 w Unii Europejskiej (Legal Foundations of the NATURA 2000 Network in the European Union). In: Furmankiewicz, M. and Mastalska-Cetera, B. (eds), *NATURA 2000 na obszarze Sudetów*. Jelenia Góra and Wrocław, Muzeum Przyrodnicze and Uniwersytet Przyrodniczy Katedra Gospodarki Przestrzennej, pp. 9–21.

Zegar, J., 2010. Warunki środowiskowe a ekonomika gospodarstw rolnych (Environmental Conditions and the Economics of Agricultural Holdings). *Wieś i Rolnictwo* 1 (146), 106–20.

Chapter 11

Linking Locally: Second Home Owners and Economic Development of the Rural Community

Krystian Heffner and Adam Czarnecki

Introduction

The location of second homes is usually connected with rural areas, but it is conditioned by different factors – both social and spatial, and also psychological and economic. At the same time, second home owners and users from outside the village are becoming an increasingly significant component of local trade and service markets and of the budgets of the rural municipalities. The degree and range of integration with the rural environment, the internalization of the needs and opportunities of the local community and the intensity of participation in rural life is of great importance (Brida, Barquet, Risso, 2010).

One of the most frequently proposed directions of economic development of rural municipalities is various forms of tourist activity. Areas of a high natural value or areas not too far from major cities are in a particularly advantageous situation. It is possible to achieve a measurable economic effect, and second homes are a specific form of individual tourist development (and progressive tourist urban development). Although the concentration of such phenomena mainly occurs around the suburban conurbations, it is gradually embracing further, often peripheral areas (counter-urbanization). These areas are characterized by significant disparities between individual and collective forms of leisure, while the first group is predominant. In other words, a relatively high degree of second-home saturation might entail a significant, positive impact on the capital resources of the municipality and on the income of its residents.

There are indications that the development of the service environment of second homes might be a potential, alternative way of social-economic activation not only for municipalities of outstanding natural value but also for other rural areas. According to research (i.e. Aledo and Mazón, 2004; Módenes and López-Colás, 2007), the main factor that determines location of second homes is proximity to large cities and transport accessibility. Thus the possibility of new job creation besides agriculture (trade and services) related to meeting the needs of second home owners is not limited to municipalities in areas of outstanding natural and tourist value. Second homes are also significant accommodation elements in many regions. They are not typical tourist facilities, but because of their non-commercial character they influence the tourist sector in rural areas. This phenomenon is occurring in virtually every municipality (to different degrees) and it involves thousands of people in Poland.

The main aim of this study was to assess the scale and nature of economic relationships (producer–consumer) between second home owners and local people, in the context of the ability to create strong long-term linkages between them. In other words, the assessment is made from the perspective of building 'short' and sustainable supply chains between the two main groups of stakeholders. It was particularly important to measure second home owners' demand for local agricultural produce and other goods and services. In addition, attempts were made to identify opportunities and obstacles to establishing and strengthening such relationships.

Literature Overview

Rural areas have become increasingly popular as a place of permanent residence among ex-urban populations and for several decades this has been particularly noticeable in the inner and outer-ring suburbs of large Polish cities. This process has continued to spread into peri-urban areas, since their unquestionable pull factors attract ex-urban residents, e.g. high amenity values, easy transportation accessibility, as well as the easy availability and convenient prices of land or empty/abandoned buildings. Nowadays, such tendencies have become more common even in some agriculturally monofunctional 'deep' rural areas, usually located far from main urban nodes (Buller and Hoggart, 1994) considered as an exemplification of current transformations of rural–urban relations, i.e. counter-urbanization, ex-urbanization, or semi-urbanization (Champion, 1989; Halfacree, 2008; Lowe and Ward, 2004; Mitchell, 2004). In general, they are based on a definite relocation of ex-urban populations and economic activity to the rural hinterland (Aledo and Mazón, 2004; Winther and Kalsø Hansen, 2006) where apart from farming, only housing and, to a lesser extent, basic services contribute to the local economic/functional structure.

Alongside the above-mentioned group of ex-urban dwellers, there has been a constantly growing urban population that lives in rural areas seasonally (e.g. second home owners and their families). They construct new, purpose-built secondary residences or, as what is particularly common in peripheral, formerly agricultural regions, they acquire abandoned farm buildings, disused agricultural land, and wastelands and then convert them into seasonal dwellings. While living permanently in cities, they own a variety of forms of second homes, which are mostly used for leisure and recreation purposes but also increasingly for work-related reasons. In developed European countries, besides traditional tourism-based communities, second homes have recently been concentrated even in some less attractive leisure destinations, e.g. agricultural or formerly agricultural areas, peripherally located and currently undergoing intensive economic restructuring and suffering from serious depopulation. According to Statistics Finland (2010), the most dynamic increase of the number of second homes has been observed in the category of 'sparsely populated' and 'rural heartland areas', while in Poland depopulation has been recognized as a leading factor influencing the spatial distribution of second-home units (Heffner, 2011). While nested in the local economy/development approach, second homes constitute one of the key forms of tourist urbanization and at the same time they provide a meaningful impulse to further socio-economic transformations of rural spaces (Galster et al., 2001). Thus second homes can apparently be considered as a stable element of the countryside and at the same time as a key determinant of rural change (a 'change factor') in contemporary/modern societies towards the post-productivist countryside (Brida, Osti, and Santifaller, 2011; Halfacree and Boyle, 1998; Marsden and Sonino, 2008).

The widespread view among tourism and rural researchers dealing with this subject is that second homes provide a meaningful/powerful boost to rural economies (Leppänen, 2003; Tamer et al., 2006). The most often mentioned include income and employment opportunities for various economic sectors (Hall and Müller, 2004; Hoogendoorn and Visser, 2011; Nielsen et al., 2009; Nordin, 1994; Sikorska-Wolak, 2007) considering that second homes create a potential demand not limited to the tourism industry but which is significantly wider, including the construction sector, sales of building materials, home equipment, technical installations and facilities as well as security, cleaning and other types of services. Given that second homes have begun to appear even in a 'deep' countryside (undiscovered by traditional mass tourism) local authorities, farmers, and entrepreneurs may consider them as an emerging development opportunity (but not as a universal remedy) contributing to the restructuring of declining rural communities or tackling/alleviating major structural and social problems, i.e. depopulation, unemployment (including hidden employment in agriculture), the limited absorption capacity of local labour markets, as well as low demand for goods and services provided by local businesses (private sector) and limited access to the local public services (Müller, 1999). Second home owners have become an increasingly important group of stakeholders with regard to the dynamics and growing diversity of local trade and service markets, resulting in the affluence of rural communities. Of great significance is the extent and strength of economic integration between newcomers (seasonal residents) and the locals. It is determined on the one hand by the ability of the local community to recognize and satisfy everyday requirements or even more particular needs of second home owners and, on the other, by the ability of the latter to express their demands and desires and their willingness to meet them locally. Strong economic linkages between second home owners/users and local people may result in an increase of farm and rural household incomes, preventing further impoverishment, improving wellbeing, contributing to functional diversification as well as to the development of technical infrastructures and improving access to public services.

Large numbers or a high concentration of second homes can constitute a particularly favourable condition (or in some cases a necessary condition) to diversify the local economy and bring the idea of multifunctional (rural) development to life. There is an economic justification/rationale behind the rural multifunctionality concept in the practice of seeing that second home owners express highly diverse demands for food products, other goods and services offered locally by farmers, craftsmen, entrepreneurs and other individuals. New demand will effectively be reflected in local people's initiatives to satisfy the needs of second home owners. In consequence, it is more likely that the financial situation of local people will be improved and stabilized through the rise in incomes and the diversification of their sources in rural Poland which is still dominated by farming or non-earned benefits, such as in the case of retired and unemployed people. However, this does not mean that second homes constitute a development opportunity exclusively or mainly for the enterprise sector – whether newly established or existing. In the Polish context, in many cases they constitute a remarkable, economically and professionally attractive opportunity for rural people to supplement their incomes, at least seasonally, and to change their occupations by providing goods and services to second home owners (Czarnecki and Heffner, 2008).

However, it seems that in Polish strategic planning second homes have been accorded far too little recognition, still being a kind of a 'middle ground' between traditional tourism (including farm-based tourism, which has been insistently recommended as an important development direction in many local strategies) and the permanent migration from urban to rural areas. It is also unexpected and difficult to explain that even some of the nationwide strategies refer only to a few and solely negative images of second homes, e.g. treating them as an obstacle to the functionality of agriculture (by reducing agricultural land), or as a factor influencing unfavourable transformations of the landscape and the cultural (local architecture) and natural environment (MRiRW, 2010). In contrast to strategic documents, local and regional policies that promote traditional mass tourism as a key driver to find a new stimulus and to diversify the local economy, second home tourism seems to be more logical/ suitable and beneficial for local community development, even if it may bring some disadvantages.

Balancing the unequal distribution of incomes from the traditional forms of tourism over the year (Butler, 1994; Ferrandino, 2005; Marcouiller and Xia, 2008; Kurtulus and Sevki, 2010) is an issue of the greatest importance for the long-term vitality of the local economy. This is reflected in the dynamic but only seasonal growth of local people's incomes, who offer goods and services to second home owners, besides more or less stable incomes for other permanent residents (farmers who develop commercial farming or daily commuters employed in manufacturing or services in nearby towns/cities). This means that it may affect and be particularly strongly felt in the economies of popular tourism regions/locations, the structure of which is usually significantly monofunctional (dependent on tourism and tourism-related economic activities). It has to be noted that seasonality is reflected not only in the differentiated income streams but, equally importantly, in obtaining and holding staff, low returns and investment (Butler, 2001), as well as in the unprofitability or even depreciation of facilities for tourists. In this context the main challenge for traditional tourism localities is to maintain a stable/comparable year-round level of income (which drastically decreases off-season) among local people, including entrepreneurs, their employees and other individuals. According to D. Marcouiller (1997: 359) 'marketing efforts often target the draw of travellers to shoulder seasons, or early/late months, to reduce these seasonal demands, but [...] some communities and regions have enjoyed limited success in expanding the tourist season'.

Given that seasonality is the prevalent characteristic of traditional models of tourism as well as some of alternative forms (e.g. farm-based tourism, eco-tourism, etc.), second homes seem to be considerably less dependent on time/holiday period fluctuations. The elderly population constitutes a large segment of second home owners, nevertheless they are less likely to be spatially mobile (which is reflected in the lower number of visits to their second homes over the year); at the same time they usually spend much more time there than people of working age. What is more, a similar effect may be achieved as long as there is a short distance between a second home and the permanent residence, good accessibility and, for properties in the mountains, the possibility of spending holidays (at least) twice a year (in the summer and winter seasons). So the demographic features of the owner as well as the spatial and geographic characteristics of the area where a second home is located, in terms of proximity, transport accessibility and winter/off-season attractiveness, may induce longer stay periods, and consequently higher local expenditure. On the other hand, based on the results of one of the Polish surveys, in the location of the second home family roots were the most potent driving force in the extension of the stay period (Heffner and Czarnecki, 2011). Thus second home owners' local purchasing practices (expenditure patterns) may provide a measurable financial effect for local suppliers. And even though it is not as significant as in typical tourism locations in the peak season, it is definitely more stable and balanced (sustainable) since there are no dramatic changes in the levels of demand and income over the year. It is basically in line with what P. Downing and M. Dower stated: 'the annual influx of second home owners into importing regions represents a flow of permanence which is not characteristic of any other form of tourism' (Downing and Dower, 1973: 29).

In light of the brief discussion above, two main disadvantages/development barriers inherent for tourism in general are not substantially attributed to second home tourism, i.e. location and seasonality/time obstacles. The spatial concentration of second homes is not limited only to well-known/popular, 'amenity-rich' or metropolitan regions, and at the same time, their use is more evenly distributed over time/the annual cycle. Consequently it is assumed that second homes may provide more development opportunities in the long-term and have a far more beneficial influence on the local/rural economy than traditional models of tourism. However, it should be mentioned that like every impact, the impact of second homes on rural communities has its positive and negative aspects. Among the downsides usually mentioned by researchers is the insignificant positive influence on job creation, i.e. a relatively small number of new jobs, their seasonal and part-time character, employment insecurity and low wages in comparison to jobs generated by other industries (Gallent et al., 2002; Luloff et al., 1994; Marcouiller, 1997; Wallace et al., 2005). What is more, second home owners' expenditure patterns (greatly influencing the degree of local economic integration and the contribution to local financial resources) are heavily dependent on travelling time and distance. Since secondary residences are very often located in suburban and peri-urban areas of large cities, and thus not far from second home owners' permanent residence, the demand for local goods and services and consequently the amount of money spent locally is relatively low. The financial leakages from the local (community) economy are thus significant and do not contribute to rural household incomes.

Research Concept and Methodology

The main objective of the study was to assess the impact of second home tourism on the local economy. In fact, the impact assessment was based on the identification of economic relationships established between second home owners and the local people (farmers, entrepreneurs, and other individuals). These linkages were then evaluated in terms of scale (number/proportion of players involved), economic value/monetary terms and the object of transaction (goods and services purchased by second home owners). In addition, attitudes and opinions expressed by local people and local authorities towards second homes, their owners and potential/future opportunities and advantages resulting from their second home tourism were also taken into account as a backdrop for the examination of the strength of economic relationships.

The study's main research method was correspondence analysis. This statistical technique is most suitable for analysing contingency tables, especially when the data are not expressed in continuous values but in categories, as was in the case with this research. This method was particularly useful in examining the relations and dependencies of the second home owners' expenditure patterns as well as local people's supply channels according to the socio-demographic characteristics of various communities and respondents.

The survey was conducted in early autumn 2009 in 20 selected communities in Poland (Figure 11.1). They were selected according to the expert method, taking into account the (highest) percentage of second homes (in relation to the total number of housing resources) as the main selection criterion (Figure 11.2), and the following additional criteria: various functional types (based on the various functional typologies of rural communities (i.e. Bański, 2009; Gwiaździńska-Goraj and Jezierska-Thole, 2013; Hełdak, 2013)) and the location of communities in one of Poland's main physiographic/landscape zones (mountains and foothills in southern Poland, lowlands in central and eastern Poland and the coastal zone in the north). Direct interviews with second home owners and local people were the main data collection technique. In total, 400 interviews were conducted with second home owners and 200 with local people. The data from direct interviews were supplemented by data based on questionnaires mailed to 610 communities (in fact, to the representatives of local authorities in communities with the highest proportion of second homes in the total housing resources (according to the official statistics)). The response rate was low, at only 15.2 per cent.

Figure 11.1   Communities surveyed and their functional type according to administrative and physiographic regions in Poland

Source: Own study, based on Kondracki, 2002 and Bański, 2009.

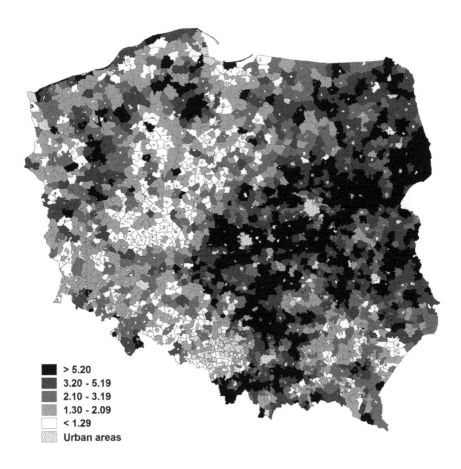

■	> 5.20
■	3.20 - 5.19
■	2.10 - 3.19
■	1.30 - 2.09
□	< 1.29
▨	Urban areas

Figure 11.2   Percentage of second homes in the total housing resources in Poland in 2002

Source: Own study.

Research Results

Second Home Owners' Perspectives

The study confirmed that the occurrence of second homes can usually determine the growth of income of the local population, mainly through the benefits from the sale of land or property. Economic impacts on local/rural community resulting from the highest concentration of second homes in the countryside, e.g. sales of food products directly from the farm, sales of food and other everyday products in local shops, sales of building materials, catering, construction services, caring for second homes out of season – housekeeping, cleaning, etc. – were clear but not very significant.

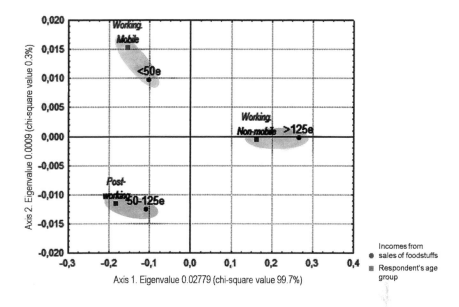

Figure 11.3   Expenditure on agricultural produce purchased directly from local people according to the respondent's age

Source: Own study.

Around 37 per cent of second home owners declared purchases of agricultural produce directly from local farmers; of these the most common group were women and 'elderly' people of post-working age. In addition, the results of the correspondence analysis showed a positive correlation between the age of the respondent and the level of demand and, in consequence, the scale of purchases (Figure 11.3). Significantly more respondents with secondary and higher education than with vocational and primary education bought such produce. However, it should be mentioned that the proportion of this population group with the lowest level of education was relatively low. The buyers of agricultural products were usually those who were maintaining their household on non-earned sources of income, mostly pensioners and, among the working population, qualified workers and specialists, while there was no clear diversification of the phenomenon in terms of the monthly net income of a household.

In addition, local products were bought more frequently by respondents whose main motivation for choosing the location their second home was the 'natural, clean environment', followed by family roots and connections with the locality – described as a respondent's previous place of residence. Obviously, agricultural produce was mostly purchased by the owners of second homes in small localities rather than in central/main villages or towns (Figure 11.4). Those surveyed usually purchased dairy products (milk, eggs, and cottage cheese) and honey directly from farms. With lifestyle changes, marked by a drive to lead a life in harmony with nature (McGranahan, 2008; Ray and Anderson, 2000) and to return to more traditional values (Hoey, 2005) (considered together as an 'amenity migration' (Gosnell and Abrams, 2009)), the pursuit of the rural idyll (Cloke and Milbourne, 1992; Rye, 2004), coupled with fashions/eco-fashions, for example a craze for organic food, urban people are tending to look for unprocessed food, produced by relatively simple methods without using chemicals. Purchasing products directly from small-scale family farms, especially if they operate in areas with high-quality natural environment, can thus be seen as an exemplification of contemporary food consumption trends.

Furthermore, some additional factors enhancing 'informal' foodstuff transactions to second home owners may be noted – primarily, strong social ties with local people, willingness to make a permanent move to a secondary residence in the future and long-term ownership (over 20 years) (Figure 11.5). It seems that not only may basic explanations resulting from respondents' socio-demographic characteristics be crucial to learning more about the origins and reasons behind the level and variety of demand for local food products, but also the 'soft' factors of becoming rooted in the village/locality. These may thus be considered as important as other most evident preconditions.

Unlike the purchases of agricultural produce, demand for informal services by local people (cleaning, mowing the lawn, raking leaves, clearing snow, looking after houses) was lower, concerning only 24 per cent of those surveyed. In the case of 'informal' services, Polish people used security/watching the house (due to the some incidences of housebreaking) and construction/renovation services. A higher percentage of people purchasing local services was also observed among the second home owners with a secondary and higher level of education than with primary and vocational education. In addition, the self-employed and pensioners also used private informal services (from the local population) more often than people directly employed did (Figure 11.6). It seems that the reason for this was elderly respondents' greater demand for help with manual/ physical work and the relatively short periods that wealthy businessmen spend in their second homes, usually 'suffering from' a lack of free time. Such services were more popular among respondents who had to commute long distances to their second homes (e.g. located on the coast and in the Kłodzko Valley). This was confirmed by the correlation between the percentage of people purchasing informal services and the distance between their permanent and seasonal homes. Along with the increasing distance (measured by travel time) the percentage of second homes owners who used such services rose from 22 per cent for those commuting less than 60 minutes to 34 per cent for those commuting over 180 minutes. Longer distances usually meant spending a shorter time or making only a few visits to a second home (especially off-season). As a result, a greater involvement of local people in providing simple but indispensable work seems to be necessary.

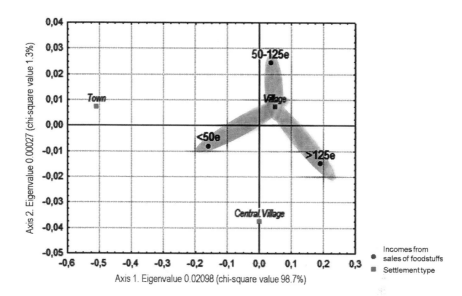

Figure 11.4   Expenditure on agricultural produce purchased directly from local people according to settlement type

Source: Own study.

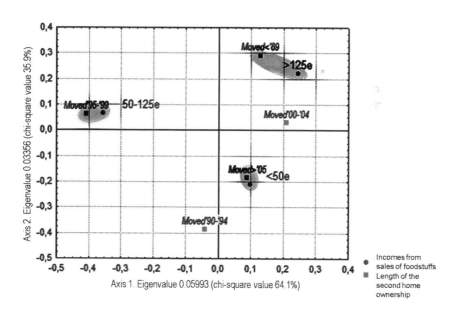

Figure 11.5   Expenditure on agricultural produce purchased directly from local people according to length of ownership

Source: Own study.

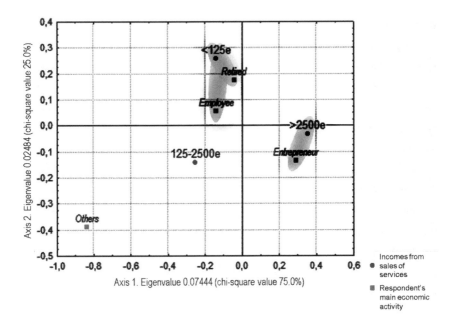

Figure 11.6   Expenditure on services purchased ('informally') directly from local people according the respondent's main economic activity

Source: Own study.

Second home owners' income was in fact one of the most important variables that differentiated the number of people purchasing informal services, and, unlike purchases of agricultural produce, higher income usually meant more frequent economic relations in terms of the use of informal services, and in consequence higher expenditure on this. Of the respondents with the lowest monthly net household income (up to €750), 17 per cent purchased informal services from the local population, compared with 22 per cent of respondents with an average income (from €750 to €1,250), and 40 per cent of respondents with the highest incomes (more than €1,250). Higher incomes meant greater purchasing power for second home owners and, at the same time, wider options in its distribution (Figure 11.7). The cost of (informal) services provided by the local population ranged from €10 to €12,500, with an average of almost €965 per purchaser per year. Larger amounts (over €500) were spent by the owners of second homes in typical tourist regions (e.g. the Kłodzko Valley in Lower Silesia and coastal area), and also by respondents with university degrees, in retirement, working as a specialists or running their own business and among the wealthiest (with a monthly net income over €2,500 per household).

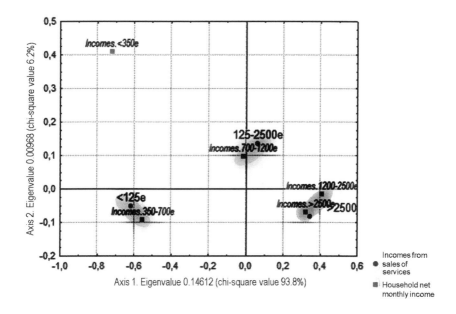

Figure 11.7   Expenditure on services purchased ('informally') directly from local people according the respondent's household monthly income
Source: Own study.

Local People's Perspectives

As regards local people's opinions on opportunities for supplying services to second homes, we can see that there was a visible prevalence of those who shared the most positive view (or positive view in general including high and very high rates), although the percentage of those who rated development opportunities most negatively was also quite significant. The main socio-demographic attributes of respondents, including age, education, and economic activity, considerably influenced their opinions (Figure 11.8). Better educated and young people had more positive views on such opportunities than older generations did. They were probably more aware of the real potential that second homes may create for local service providers. This positive view was widespread among local entrepreneurs, not only those selling services to second home owners, but especially those who earned the most. On the other hand, a lack of or poor opportunities were seen mainly by less-educated and elderly people, and by owners of commercial farms (producing mainly for the market). It seems that for large-scale (commercial) farmers second home owners' demand for agricultural produce/foodstuffs was too low (marginal/limited) to take it into account in economic terms. Their mostly negative opinions may also be affected by their quite common belief and negative feelings that urban people (including seasonal residents such as second home owners) or external investors tend to buy out agricultural land for other purposes, and consequently limit their opportunities to develop their main farming activity.

A particularly positive view on service-supply opportunities resulting from second homes was shared mainly by those who lived in multifunctional and tourism-based communities (especially those on the coast) as well as those from small towns and large villages (local service centres) (Figure 11.9). It seems that respondents indirectly highlighted localities with the most beneficial conditions for starting and developing economic activities – high demand for services (shown not only by second home owners, but other consumers, including traditional tourists). In formulating their opinions, among other advantages they might also take into consideration the well-developed technical infrastructure (good accessibility, transport system) and other business facilities/environment as well as a high concentration and great variety of economic activity in general.

In contrast, some negative opinions were expressed by respondents from typically agricultural communities and those near large cities. Although these communities represented two distinctive functional types, which differ significantly from each other, in both cases the main obstacle to developing services was probably the low level of demand for them by second home owners.

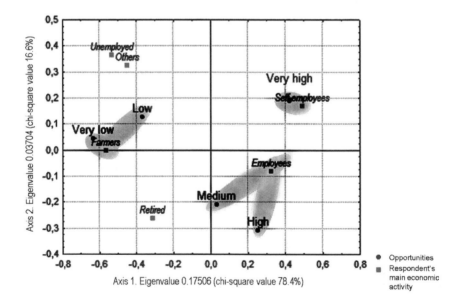

Figure 11.8   Opportunities resulting from second homes according to the respondent's main economic activity

Source: Own study.

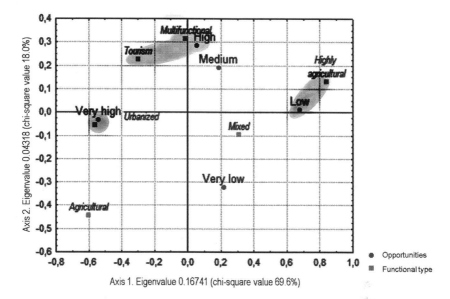

Figure 11.9   Opportunities resulting from second homes according to the functional type of community

Source: Own study.

There are not as many second homes in agricultural communities as there are in tourism-based communities and they are spatially dispersed, as rural settlements usually are in Poland. In suburban and peri-urban communities, although there is a high concentration of second homes and a large number of permanent residences, the level of real demand is also quite low. This seems to be unreasonable at first sight, but it may be largely explained by the expenditure patterns of second home owners. Very often they are more likely to purchase goods and services in their permanent place of residence, usually a nearby city, because of easy access to the shops (the retail market in general), cheaper prices, and the wider range of products and services.

Based on local people's opinions, opportunities to supply seasonal residents with services were significantly higher in traditional second home and leisure communities. This positive feedback came particularly from respondents in communities where second homes began to appear in the 1980s or even earlier. On the other hand it is interesting that local people from communities where second homes were quite a new phenomenon did not consider them as a business opportunity, even though they may have seen a lot of new residences being built after the year 2000, which required great support from construction companies and renovation and infrastructure facilities.

Besides local people's direct opinions on the development opportunities resulting from second homes for local services, there is also a possibility of understanding this issue better in a more indirect way – based on the data on existing economic relationships with second home owners. Only 12 per cent of local people (private individuals) provided second home owners with services using 'informal' channels (without being officially registered as a firm/enterprise). In fact, this is a relatively small proportion of service providers in comparison to the over 24 per cent of second home owners who admitted buying services directly from local people. It seems that the differences may be accounted for by two main factors: (1) the relatively small number of farmers supplying a relatively large number of second home owners; or, which is probably more important to explain the whole situation, that (2) the trade exchange between second home owners and the locals is usually outside the official and traditional channels, i.e. unofficially, without tax. As a result, local people from traditionally agricultural areas involved in this kind of economic relationship were rarely or never willing to speak frankly about their role in the practice or the amount of money they earned.

As regards the distribution of local people's earnings from the sale of agricultural produce to second home owners, based on the results of the correspondence analysis the following explanatory factors may be mentioned: community landscape type, respondents' age, his/her main economic activity (Figure 11.10), as well as local people's opinions and attitudes towards second homes in their village/ community (including the possible inconvenience resulting from second homes and the influence of second home owners on the local infrastructure facilities).

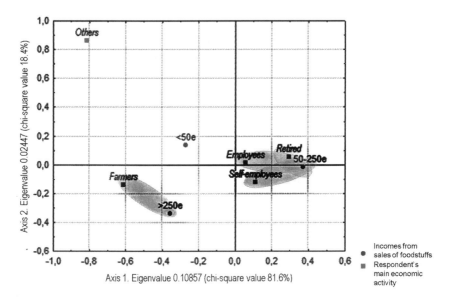

Figure 11.10 Distribution of incomes from sales of agricultural produce according to the main economic activity

Source: Own study.

Generally, considerably higher earners permanent residents of the mountain regions, mostly farmers, people of a mobile working age with generally positive view of the second home phenomenon with regard to community development and the impact on infrastructure facilities.

In the case of services provided 'informally' to second home owners, local income distribution was mostly determined by the community landscape type (with the highest earnings on the coast) and the basic respondent's socio-demographic attributes (including age, sex, education, and main economic activity). Those providing 'unofficial' services were the younger generation, men, with a secondary or basic vocational education, as well as people running their own businesses (except farming) and employees (Figure 11.11). This was mainly due to the fact that some activities, such as house building, renovation and repairs, require the relevant professional qualifications and training. On the other hand, services that do not require either much physical efforts or investment, like minding and looking after the house, or services that are relatively simple in themselves, such as cleaning the property, were more often provided by the older generation and people whose main or only official source of income is social security benefits (retired people, the unemployed and even disabled people, of course depending on the level of disability).

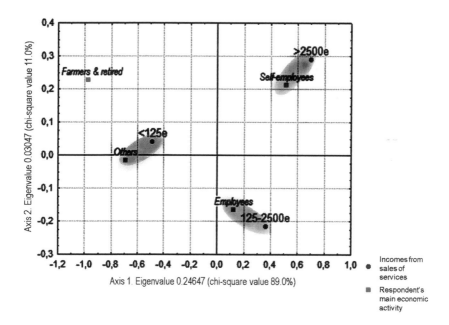

Figure 11.11 Distribution of incomes from sales of services according to the respondent's main economic activity

Source: Own study.

Local Authorities' Perspectives

Besides economic relationships between seasonal residents and the locals, it was shown that second home tourism has clear but more indirect implications for the local economy. By paying property taxes and other local payments, second home owners contribute to municipal financial resources, i.e. increasing incomes to the local council budgets as well as diversifying the income structure.

According to the majority of respondents (i.e. representatives of local authorities, including mayors, officials and inspectors for the development of the municipality, external fundraising, tourism, planning, and zoning), one can note a variety of positive aspects for community development resulting from second home tourism. First and foremost, respondents mentioned that second home owners brought 'concern about cleanliness, order, and security' to the village (Table 11.1). Furthermore, among the advantages of second home tourism the officials surveyed singled out the 'preferable spatial development of the village', usually considered to be an improvement of the infrastructure for tourists, reflected in the number of newly established tourist services (catering and retail businesses, as well as in the renewal or construction of bicycle paths, waterways, signage, etc.), in the improvement of the quality of local roads as well as in the maintenance/restoration of the local/traditional architectural style. Additionally, respondents mentioned other positive results for community development resulting from second homes, including housing density, so that expanding the infrastructure (roads, waterlines, gas supply system) becomes less complex technically and economically more feasible; and the use of low-quality agricultural land for leisure and recreation purposes or even the improvement of the rural spatial design/planning.

Regarding the benefits for the local economy, approximately a quarter of the local authority representatives highlighted tax revenues (mainly from properties and land) as one of the most positive effects of second home tourism in the community. The reasons for not recognizing their role in this context (especially in typical tourist communities) should be seen in the difficulty of estimating the amount of revenue collected (in Poland most municipal offices do not keep records for this) or its low share in the structure of local community income. However, it seems that this component of municipal budgets is understated by the local authority representatives and consequently also underestimated. Naturally, the amount of income not only depends on the size of the house, land or the number of second homes (which in suburban communities are often 'second' only in name, while in fact they are used all year round by owners who are not registered there as permanent residents), but also on the rate per square meter set by the local authorities.

According to experts from *Gazeta Prawna* rates may range from €0.15 per m^2 (in which case the rate is the same as for a permanent/first home), up to €1.60.[1]

---

1   Information from *Gazeta Prawna*, 24 June 2008 (no. 122) – 'Za domek letniskowy podatek wyższy niż za willę' ('Tax higher for a holiday home than for a villa').

Table 11.1 **Local authority opinions on the advantages for community development resulting from second homes according to community characteristics**

		High care of cleanliness and security		Better spatial development/ order		New jobs		Tax revenues	
N		N	%	N	%	N	%	N	%
Total		60	65.2	45	48.9	28	30.4	23	25.0
Community functional type	Urbanized	3	33.3	6	66.7	1	10.0	2	20.0
	Multifunctional transitional	8	61.5	4	30.8	5	38.5	4	30.8
	Highly agricultural	10	100.0	4	40.0	3	30.0	1	10.0
	Prevalence of agricultural function	18	69.2	13	50.0	5	19.2	6	23.1
	Tourism-related and recreational functions	5	45.5	6	54.6	7	63.6	3	27.3
	Mixed functions	15	71.4	11	52.4	7	33.3	7	33.3
Administrative status	Urban	2	66.7	2	66.7	0	0.0	0	0.0
	Rural–urban (semi-urban)	19	57.6	16	48.5	12	36.4	11	33.3
	Rural	39	69.6	27	48.2	16	28.6	12	21.4
Number of permanent inhabitants	≤4,999	22	71.0	19	61.3	8	25.8	8	25.8
	5,000–9,999	24	70.6	15	44.1	10	29.4	9	26.5
	≥10,000	14	51.9	11	40.7	10	37.0	6	22.2
Number of secondary residences	≤49	13	81.3	7	43.8	7	43.8	4	25.0
	50–99	24	77.4	15	48.4	9	29.0	5	16.1
	100–199	17	58.6	15	51.7	9	31.0	9	31.0
	≥200	6	37.5	8	50.0	4	18.8	5	31.3
Percentage of secondary residences	≤4.9%	25	69.4	14	38.9	14	38.9	8	22.2
	5.0–9.9%	23	62.2	19	51.4	10	27.0	10	27.0
	≥10.0%	12	63.2	12	63.2	4	21.0	5	26.3
Community tourist attractiveness assessment	Low	15	62.5	9	37.5	6	25.0	4	16.7
	Average	31	66.0	23	48.9	13	27.7	11	23.4
	High	14	66.7	13	61.9	9	42.9	8	38.1

Source: Own calculations.

However, in reality rates applied by local governments rarely come close to the upper limit. If we assume the lowest rate, and the average house floor space ranges from 50 to 100 m² (based on the results of the survey), then the diverse number of second homes (based on the 2002 census) would bring annual local budget revenues of €103 to €207 in the Bieszczady municipality of Lutowiska (with the lowest number of second homes in the country) to €11,867 to €23,733 in Zgierz (with the largest number of second homes), and an average of €1,076 to €2,151. However, assuming the maximum rate, the numbers would range from €1,115 to €2,230 in the municipality with the lowest number of second homes to €128,161 to €256,233 in the one with the highest number, with an average in the range of €11,609 to €23,219. In these cases, income from property tax would account for 0.02 per cent to 0.5 per cent of the annual local budget of Lutowiska and 0.54 per cent to even 11.7 per cent for Zgierz. Though not always significant, in some cases at least it means a considerable stream of funding.

Other positive effects of the impact of second homes on the local economy were: job creation (non-agricultural employment), farm diversification, and contribution to the local council budgets (property tax payments). Respondents saw the second home phenomenon as a factor generating new workplaces/jobs in rural labour markets, whether by initiating their own business or as a manifestation of employment growth in existing local enterprises as a consequence of second home owners' increasing demand for some local goods and services.

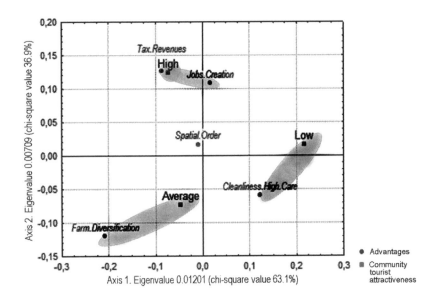

Figure 11.12 Advantages resulting from second homes according to municipal tourist attractiveness

Source: Own study.

What is more, second home tourism as a factor generating jobs was also the most common consideration among local authority representatives, who regarded their municipal natural and human assets as highly attractive for 'tourism and recreation' activities, compared to those assessed as less attractive (Figure 11.12). In addition, according to the respondents enterprises providing tourist services to seasonal residents play a leading role in meeting the demand from second homes users, and they are the greatest beneficiaries of the further development and concentration of second homes. Interestingly, researchers from the Nordic countries share an almost common opinion about the negative impact of the phenomenon of second homes on the development of the tourism industry, as due to their specificity they stimulate only certain economic activities, other than those stimulated by traditional branches of tourism – accommodation services, transport, catering, recreation, etc. (Müller, 2002).

Construction–renovation services were mentioned by local authorities as the second important industry sector that benefits most from second homes. These services have been recognized in the scientific literature as one of the basic/core activities developed from second homes (new construction, renovation or adaptation of their former farm buildings for second homes – converted second homes), and as a third important factor – retail trade. Interestingly, it proves that to some extent respondents' opinions on economic opportunities resulting from second homes were influenced by their perception of the local/community natural and human attractiveness for tourists. The greater attractiveness, the more often benefits and advantages were mentioned for local economic development (mainly income and employment growth).

The changing economic profile of farms (farm diversification) was considered as a positive result of second homes by more than one fifth of respondents, who highlighted organic farming and horticulture as a new/supplementary activity. Farms' adaptation/adjustment to the new circumstances (the specific demands of seasonal residents for fruit and vegetables) by developing organic farming seems to be the most obvious and logical, especially in light of the currently observed pattern in modern urban societies for promoting and disseminating 'healthy lifestyles', 'clean' looking eco-agricultural products or those produced by traditional methods. Interestingly, the beneficial diversification of farm activities was mainly mentioned by respondents from the communities with the greatest population potential, highly developed tourism and recreation, and the most touristically attractive. This means that the phenomenon of second homes is not an impulse to start the process of diversifying the local economy, but it is a factor that stimulates further functional transformations in existing multifunctional structures/units.

Conclusions

The spread of the second home phenomenon into rural areas previously outside of the mainstream interest of metropolitan (or in general urban) population or tourism and leisure centres, makes it an important issue affecting the economic, social and spatial development of rural areas. It has been of particular importance in Poland in the period of joining the market-economy system and since accession to the European Union. One of the indirect effects of EU enlargement may be a significant increase in the interest in investing in second and vacation homes for foreigners who often live in more remote rural areas that are usually considered less attractive. In the demographic, socio-cultural and economic contexts, the gradual evolution of second homes into places of permanent residence becomes significant (a process that accords with the phenomenon of suburbanization and counter urbanization).

The research has made it possible to identify various aspects of the impact of second homes on the development opportunities for rural areas and, in particular, villages in which they were located. Influence was generally reported:

- in the improvement of provision and access to rural basic services in the locations of second homes (electricity, water supply, sewage system, road network, healthcare centres, etc.) as well as of the development of facilities for tourism (e.g., bike paths, hiking trails etc.);
- in the emergence of new morphological (settlement) structures in rural areas and their spatial development (new enclaves or even entire settlements/ communities of second or vacation homes, maintenance and restoration of typical, local, or regional architectural styles);
- in slowing the depopulation process in rural areas (when a second home becomes a primary residence);
- in the increase in wealth of the rural population (additional income opportunities from the sale of land and properties, agricultural and every-day use produce, and also formal as well as informal services);
- in the noticeable increase of incomes to the local authority budgets (tax revenues and other local charges).

Among the positive effects of second home tourism we should therefore mention the economic revival of rural areas and the involvement of local structures (local authorities, entrepreneurs, farmers, and other people) in the processes of stimulation and improvement of the tourism potential in the main second home locations. In the context of the favourable influences of such phenomena on the development of rural areas, the thesis of the modernization impact of second homes on the local/rural community is justified.

In the areas of high concentration of second homes there have also been negative features in terms of the environmental, cultural and landscape context. These mostly concerned the progressive degradation of the natural environment and the spontaneous character of urbanization in rural areas (in a broad sense). Very often while expecting peace and quiet, second home owners experience an 'urban buzz', a side-effect of dynamic growth in the number of vacation/ seasonal residences in the most attractive rural areas. However, studies carried out in Poland within the scientific project 'The significance of second homes in rural development' clearly indicated that the positive effects of the location of second homes in the countryside outweighed the disadvantages, particularly in the economic sphere.

Based only on local people's opinions on development opportunities for local services, the general view is positive and second homes may be a significant stimulus for the rural economy. At the same time, the more interesting or even striking feature is that this common positive view is not reflected (or in fact only to a smaller extent) in the reality, that is in actions performed or rather not performed by local people – proved by the number of people involved and amount of money earned. The permanent local residents are very often not willing to speak frankly about sales of goods and services to second home owners, mostly because of their unofficial character. This is probably the main reason behind the underestimate of economic relationships or even more, behind the underrating of the actual significance/impact of second homes on rural community development. However, in referring to second home owners' opinions on local purchases (which seem to be more reliable), it is still not enough to prove that opinions on development opportunities have clear implications in reality or in action.

There is a need to learn more about the reasons why local people do not widely/commonly use the opportunity to sell local products or provide second home owners with services. It may be supposed that possible reasons include low demand as a result of lack of interest in purchasing local products or the low number of second home owners. From the second home owner's perspective, various reasons for not purchasing local foodstuffs include: high prices and the poor variety of products as well as difficulties in finding local producers and service providers. The latter may be definitely considered as something simple that local people could do to attract possible consumers without serious expense. However, at the same time, for many local producers to be visible it would be necessary to move away from the shadow economy, to register and run their own businesses officially. This would probably be unacceptable at least for a certain group of local producers and service providers.

Generally, due to their local frequency second homes have become a vital component of social relations in the countryside in many parts of Poland. On the other hand, economic relations where the phenomenon of second homes has a wider dimension are evident everywhere. However, due to their irregularity they are playing an increasing role that is still of small importance to the rural economy. Second homes are becoming permanent and significant element of rural areas, but their impact is less on social than on economic relations.

Acknowledgement

This work was supported by the Polish Ministry of Sciences and Higher Education under the grant 'Significance of second homes in the development of rural areas'

(N N114 122935) and by the Academy of Finland under the grant 'Homes beyond homes – Multiple dwelling and everyday living in leisure spaces (HOBO)' (SA 255424).

References

Aledo, A. and Mazón, T., 2004. Impact of Residential Tourism and the Destination Life Cycle Theory. In: Pineda, F.D., Trebbia, C.A., and Mugica, M. (eds), *Sustainable Tourism*. Southampton, WIT Press, pp. 25–36.

Bański, J., 2009. *Typy obszarów funkcjonalnych w Polsce*. Instytut Geografii i Przestrzennego Zagospodarowania PAN, Warszawa, (expert assessment). Available at: http://www.igipz.pan.pl/tl_files/igipz/ZGWiRL/Projekty/Ekspert yza_typologia.pdf. Accessed: November 2010.

Brida, J.G., Barquet, A., and Risso, W.A., 2010. Causality Between Economic Growth and Tourism Expansion: Empirical Evidence from Trentino-Alto Adige. *Tourismos* 5(2), 87–98.

Brida, J.G., Osti, L., and Santifaller, E., 2011. Second Homes and the Need for Policy Planning. *Tourismos: An International Multidisciplinary Journal of Tourism* 6 (1), 141–63, UDC: 338.48+640(050).

Buller, H. and Hoggart, K., 1994. *International Counter Urbanization: British Migrants in Rural France*. Aldershot, Avebury.

Butler, R.W., 1994. Seasonality in Tourism: Issues and Implications. In: Seaton, A. (ed.), *Tourism: The State of the Art*. Chichester, Wiley, pp. 332–9.

——— (2001). Seasonality in Tourism: Issues and Implications. In: Baum, T. and Lundtorp, S. (eds), *Seasonality in Tourism*. Elsevier Ltd, pp. 5–21.

Champion, T., 1989. *Counterurbanisation: The Changing Pace and Nature of Population Deconcentration*. London, Edward Arnold.

Cloke, P. and Milbourne, P., 1992. Deprivation and Lifestyles in Rural Wales. 2. Rurality and the Cultural Dimension. *Journal of Rural Studies* 8 (4), 359–71.

Czarnecki, A. and Heffner, K., 2008. Drugie domy a zrównoważony rozwój obszarów wiejskich. *Wieś i Rolnictwo* 2, 29–46.

Downing, P. and Dower, M., 1977. *Second Homes in Scotland*. Dartington, Dartington Amenity Research Trust.

Ferrandino, V., 2005. Dal turismo di élite al turismo di massa. Spunti di riflessione sulla realtá sannita. *Rivista di storia finanziaria* 15, 7–21.

Galster, G., Hanson, R., Ratcliffe, M.R., et al., 2001. Wrestling Sprawl to the Ground: Defining and Measuring an Elusive Concept. *Housing Policy Debate* 12 (4).

Gallent, N., Mace, A., and Tewdwr-Jones, M., 2002. *Second Homes in Rural Areas of England*. Wetherby, The Countryside Agency.

Gosnell, H. and Abrams, J., 2009. Amenity Migration: Diverse Conceptualizations of Drivers, Socioeconomic Dimensions, and Emerging Challenges. *GeoJournal* DOI 10.1007/s10708-009-9295-4.

Gwiazdzińska-Goraj, M. and Jezierska-Thole, A., 2013. Functional Changes of the Rural Areas in Poland. Case Study: Warmińsko-Mazurskie Voivodeship. *Journal of Settlements and Spatial Planning* 4 (1), 53–8. Available at: http:// jssp.reviste.ubbcluj.ro. Accessed: October 2013.

Halfacree, K., 2008. To Revitalise Counter Urbanisation Research? Recognising an International and Fuller Picture. *Population, Space and Place* 14, 479–95.

Halfacree K. and Boyle, P., 1998. Migration, Rurality and the Post-Productivist Countryside. In: Boyle, P. and Halfacree, K. (eds), *Migration into Rural Areas*. Chichester, Wiley, pp. 1–12.

Hall, C.M. and Müller, D.K. (eds), 2004. *Tourism, Mobility and Second Homes*. Clevedon, Channel View Publications.

Heffner, K., 2011. Zakres i zróżnicowanie przestrzenne zjawiska drugich domów. In: Heffner, K. and Czarnecki, A. (eds), *Drugie domy w rozwoju obszarów wiejskich*. Warsaw, Institute of Rural and Agricultural Development of the Polish Academy of Sciences, pp. 69–87.

Heffner, K. and Czarnecki, A., 2011. Wpływ zjawiska drugich domów na rozwój obszarów wiejskich. In: Heffner, K. and Czarnecki, A. (eds), *Drugie domy w rozwoju obszarów wiejskich*. Warsaw, Institute of Rural and Agricultural Development of the Polish Academy of Sciences, pp. 131–62.

Hełdak, M., 2013. Functional Standartization of Rural Areas of Dolnośląskie Voivodeship. *Bulletin of Geography. Socio-Economic Series* 13, 127–37. Available at: http:// wydawnictwoumk.pl/czasopisma/index.php/BGSS/ article/view/v10089-010-0010-5 [accessed: 25 May 2014], doi: http://dx.doi. org/10.2478/v10089-010-0010-5.

Hoey, B.A., 2005. From Pi to Pie: Moral Narratives of Noneconomic Migration and Starting Over in the Postindustrial Midwest. *Journal of Contemporary Ethnography* 34 (5), 586–624.

Hoogendoorn, G. and Visser, G., 2011. Economic Development Through Second Home Development: Evidence from South Africa. *Tijdschrift voor Economische en Sociale Geografie* 102 (3), 275–89.

Kondracki, J.A., 2002. *Geografia regionalna Polski*. Warsaw, PWN Publishing.

Kurtulus, K. and Sevki, U., 2010. Measuring the Seasonality in Tourism with the Comparison of Different Methods. *EuroMed Journal of Business* 5 (2), 191–214.

Leppänen, J., 2003. Finlands Skärgårdsprogram och Fritidshusboende 2003–2006. Paper presented at Second Home Ownership and Shore Line Protection: Seminar on Regional Development in the Kvarken Region, Umeå, Sweden, 25 June 2003, Helsinki, Skärgårdsdelegationen.

Lowe, P. and Ward, N., 2009. England's Rural Futures: A Socio-Geographical Approach to Scenarios Analysis. *Regional Studies* 43 (10), 1319–32.

Luloff, A.E., Bridger, J.C., Graefe, A.R., et al., 1994. Assessing Rural Tourism Efforts in the United States. *Annals of Tourism Research* 21, 46–64.

Marcouiller, D.W., 1997. Toward Integrative Tourism Planning in Rural America. *Journal of Planning Literature* 11 (3), 337–57.

Marcouiller, D.W. and Xianli, X., 2008. Distribution of Income from Tourism-Sensitive Employment. *Tourism Economics* 14 (3), 545–65.

Marsden, T. and Sonino, R., 2008. Rural Development and the Regional State: Denying Multifunctional Agriculture in the UK. *Journal of Rural Studies* 24, 422–31.

McGranahan, D.A., 2008. Landscape Influence on Recent Rural Migration in the U.S. *Landscape and Urban Planning* 85, 228–40.

Ministry of Agriculture and Rural Development, 2010. *Kierunki rozwoju obszarów wiejskich: Założenia do 'Strategii Zrównoważonego Rozwoju Wsi i Rolnictwa'*. Warsaw, Ministry of Agriculture and Rural Development.

Mitchell, C.J.A., 2004. Making Sense of Counterurbanization. *Journal of Rural Studies* 20, 15–34.

Módenes, J.A. and López-Colás, J., 2007. Second Homes and Compact Cities in Spain: Two Elements of the Same System? *Tijdschrift voor Economische en Sociale Geografie* 98 (3), 325–35.

Müller, D.K., 1999. German Second Home Owners in the Swedish Countryside. Gerum kulturgeografi 2, Department of Social and Economic Geography, Umeå, Umeå University.

Nielsen, N.C., Salling Kromann, D., Kjeldsen, Ch., et al., 2009. Second Homes: A Possible Pathway to Rural Development? Paper presented to the WorkGroup on Multifunctional Landscapes at the ESRS Conference, 17–20 August 2009, Vaasa.

Nordin, U., 1994. *Fritidsbebyggelse för Skärgårdsbor? – Studier av Fritidsboendets Betydelse för Sysselsättningen i Blidö Församling, Norrtälje Kommun 1945–1987*. Stockholm, Stockholms Universitet.

Ray, P.H. and Anderson, S.R., 2000. *The Cultural Creatives: How Fifty Million People are Changing the World*. New York, Harmony Books.

Rye, J.F., 2004. Constructing the Countryside: Differences in Teenagers' Images of the Rural. Paper presented at the XI World Congress in Rural Sociology Trondheim – Norway, 25–30 July 2004. Working Group 24: 'Back to the land' in the 21st Century.

Sikorska-Wolak, I., 2007. Społeczno-ekonomiczne przesłanki kształtowania funkcji turystycznych obszarów wiejskich. In: Sikorska-Wolak, I. (ed.), *Turystyka w rozwoju obszarów wiejskich*. Warsaw, Warsaw University of Life Sciences Publishing, pp. 13–28.

Tamer-Görer, N., Erdođanaras, F., Güzey, Ö., et al., 2006. *Effects of Second Home Development by Foreign Retirement Migration in Turkey*. Paper presented at 42nd ISoCaRP Congress, Istanbul.

Wallace, A., Bevan, M., Croucher, K., et al., 2005. *The Impact of Empty, Second and Holiday Homes on the Sustainability of Rural Communities: A Systematic Literature Review*. York, Centre for Housing Policy.

Winther, L. and Kalsø Hansen, H., 2006. The Economic Geographies of the Outer City: Industrial Dynamics and Imaginary Spaces of Location in Copenhagen. *European Planning Studies* 14, 1387–406.

Chapter 12

Local Governmental Quality and the Performance of Medium and Large Companies Across Rural Vietnam

Kai Mausch and Javier Revilla Diez

Introduction

Economic theory predicts convergence, and between 1970 and 2000 inequality across countries did decline. Yet, worldwide within-country inequality has risen (Sala-i-Martin, 2006), especially in developing countries (Schätzl, 2003; Shankar and Shah, 2001; Venables, 2003; World Bank, 2003). The extreme rural/urban differences are one main cause for this problem as the majority of poor people live in rural areas of developing countries. The share of poor people living and/or working in rural areas is estimated at 75 per cent and reached as high as 92.3 per cent in 2002 for Vietnam (United Nations, 2007), stressing the importance of the urban/rural divide. Therefore, effective strategies to overcome extreme poverty all over the world need to consider the particular characteristics of rural areas to adequately meet the needs of the rural poor (United Nations, 2005).

In developing countries in particular, agricultural production is constantly endangered by adverse environmental conditions. Additionally, high reliance on subsistence as well as few marketed products makes households especially vulnerable (Dercon, 1996; Dercon and Krishnan, 1996; Reardon, 1997). One prominent and widely discussed path out of poverty is income diversification of farmers into the rural non-farm economy (RNFE) via wage- or self-employment. The connection between the poverty status of households and their participation in the RNFE has been the subject of multiple theoretical and empirical studies over the last decade. These mainly concluded that participation in the non-farm economy contributes to poverty reduction in developing countries. Furthermore, the promotion of wage-employment in remote and low income areas is a successful strategy in overcoming income inequalities within a country (Collier, 2007; Schupp, 2002).

Considering conducive institutions as a key factor explaining economic success across economies became commonly accepted (Gagliardi, 2008) and thus the Millennium Declaration of the United Nations General Assembly states that institutional reforms promoting good governance are one central condition for the goal of poverty eradication (United Nations, 2000). Among other measures, pro-poor policies should strengthen rural information systems and develop rural institutional capacities. A good institutional environment induces investments and increases income growth. Furthermore, high quality institutions distribute income more equally by restricting the ability of the rich to engage in successful rent seeking (Gradstein, 2007).

Vietnam witnessed a strong and steady economic growth of around 8 per cent per year over the last decade. The major driving force in the Vietnamese success in recent years can be attributed to the performance of the industry, manufacturing and service sector, which accomplished growth rates of 10.2 per cent, 12.5 per cent, and 8.5 per cent respectively (World Bank, 2007).

Vietnam's business environment in general has undergone tremendous changes since the start of the doi moi[1] reform in 1986 when the socialist party introduced a more market-oriented approach (Revilla Diez, 1995). In the following decade, Vietnam witnessed the highest economic growth rate out of the 40 poorest countries in the world. Between 1992 and 1998, real income rose by 39 per cent (Fforde, 2001).

Alongside the change towards an open and world market integrated economy, the share of agricultural output declined from 43 per cent in 1980 to 24.6 per cent in 2002, whereas the share of the service sector rose from 33.7 per cent to 39.5 per cent and the industry sector rose from 23.3 per cent to 35.9 per cent (United Nations, 2007).

This process was not limited to urban areas but also strengthened the RNFE, which in 2004 accounted for over 30 per cent of rural employment (ADB, 2005). The constraints that firms in rural areas face are different from the ones located in the country's centres such as Ho Chi Minh City, Hanoi, or Da Nang. Besides the obvious constraints that rural enterprises in many developing countries face like lack of electricity, transportation and telecommunication, they also include institutional constraints such as political/regulatory uncertainty or corruption (ADB, 2005). However, the fact that land prices in urban areas are about 500 times higher than in rural areas (ADB, 2005) increases the incentives for companies to invest in rural areas.

Institutional Economics and Transition Economies: Theory and Recent Evidence from Vietnam

The most frequently applied approach in the analysis of institutions is transaction cost economics originating from Coase (1937). Coase focused on the role of exchange costs with firms viewed as governing structures with their main purpose being to organize the production and exchange process. 'Firm-level decisions, especially those regarding internal production and contracting out […], business decisions in terms of choice of activities under vertical and horizontal integration of business activities […] have been assessed in terms of incidence of their relative transaction costs in addition to traditional direct costs' (Rao, 2003: 7).

---

[1] Doi moi is Vietnamese for regeneration.

In the context of provincial differences in the performance and the development possibilities of rural non-farm companies, all categories of transaction costs, i.e. market-based, administrative/managerial and political transaction costs, are important as they all determine the actions of the entrepreneurs. As transaction costs are difficult to measure directly, they are typically included using different proxies such as the number of contracting partners, which will affect transaction costs based on the number of people dealt with, or connection to the internet and telephone network as this should lower the time spent in getting in contact with current and future contracting partners. Furthermore, the length of the relationship with market partners reduces the transaction costs, as generally the longer two partners work with each other the higher the level of trust, resulting in a lowering of the monitoring effort and therefore the transaction costs. The same argument holds for reputation, which can be signalled by holding certificates. If a company holds a specific certificate it signals that this company operates according to the rules specified by the issuing authority and therefore any new market partner can rely on this signal (Rao, 2003).

Furthermore, internal administrative processes influence transaction costs. These costs occur due to decision-making and hierarchical structures. The longer the hierarchy, the higher the transaction costs. A proxy for this kind of transaction cost is the share of professional staff as their main task is to monitor operations and organize the production and exchange processes.

As for the political factors, the general ease of doing business can be measured, e.g., by the reliability of regulations with adjustments taking time or the business attitude of the authorities taking less time to deal with officials who try to help (Rao, 2003).

Taking a look at the special case of transition economies, it becomes clear that institutions play a vital role in the transition process. After privatization, state owned companies were often split into many small enterprises that make decisions autonomously. Thus, the transition, if not accompanied by an adequate change in the institutional setting, could be trapped in several hold-ups (Cungo et al., 2008).

New institutions should not aim to outrival existing institutions but provide alternatives and thus the necessary flexibility. Private screening and monitoring facilities like trade unions and business associations serve as a complement to the legal system and thus broaden the opportunities of the companies involved (Fafchamps, 2004).

Another problem emerging from transition are increasing information costs as the former centrally distributed goods and centrally planned prices are now variable and determined by market conditions (Krug, 1991). The importance of networks for information- or risk-sharing is generally agreed upon. Fafchamps (2004) provides a unique framework for analysing how networks influence the economic behaviour of firms under non-convex transaction costs. Non-convex transaction costs refer to a situation in which economic transactions recur over time between the same individuals or firms. This leads directly to the observation that especially in former centrally planned economies the former elites have a comparative advantage.

They used to have most of the relevant market information and thus have a comparative advantage in information gathering after privatization (Krug, 1991).

The implementation and especially enforcement of these regulations and laws is a crucial part of the process according to institutional literature. For example, an anti-corruption law is in place in Vietnam but corruption levels are still high. According to Transparency International (2008), Vietnam still ranks 121st out of 180 countries surveyed. The same holds for access to land rights. Even with land being officially obtainable, private companies still face major problems in getting access to land (ADB, 2005).

As shown above, all factors introduced in the transaction cost theory are even more important in the transition process, as in the first years after a political change, transaction costs are likely to increase and need to be lowered in order to supplement growth.

Although institutional economics mainly focuses on cross-country differences, recent research showed that, even in economies which are mainly centrally planned like Vietnam, institutions also differ within countries (Malesky and Ray, 2007).

The Provincial Competitiveness Index (PCI) assessed the differences in the private sector performance among the 64 provinces of Vietnam (Malesky, 2007). According to the PCI, the worst performing areas are concentrated in the northern uplands on the border with China and Laos, the south-central Highlands around the province of Dak Lak and on the border with Cambodia. The best performance is found mainly in the Mekong River Delta around Ho Chi Minh City.

However, many of the indicators used, e.g. entry costs and access to land, are mainly relevant for establishing new businesses and less important to established companies. Furthermore, the design of the mail-out survey has some weaknesses, especially when faced with response rates of less than 15 per cent. This rate is generally high considering the design, but nevertheless is prone to severe biases. Additional concerns arise from the indicators being aggregated using weights. Finally, the PCI gained huge public attention, which leads provincial governments to try to improve their position. This possibly results in attempts to influence the result of the PCI as actual changes would take a long time to show effects. Particularly against the background of the differences in the performance of some regions from 2005 to 2007, this might have been the case. Some provinces, e.g. in the area around Hanoi, which were ranked at the bottom of all provinces have improved their scores remarkably, being rated among the best after only two years.

Objectives and Data

Objectives

Arguably, much of the institutional setting in terms of regulations and laws are set at a national level and thus are similar to all companies within a country in theory. Yet, in practice, this is often determined by the regional and local authorities. The officials who interpret the laws and apply/enforce them either stringently or more loosely often do make a difference. Another advantage of analysing the institutional setting or the governmental quality within a country rather than across countries is that a micro-level study will cover institutions better as parameter heterogeneity, unobserved heterogeneity and endogeneity do not need to be controlled for, which can cause problems in a cross-country comparison (Grimm and Klasen, 2007).

The objective of this chapter is to investigate the differences in the institutional settings, i.e. infrastructure, governmental quality, and transaction costs, of three remote provinces in Vietnam. The possible differences will be analysed in terms of their effects on economic performance and employment generation of the companies operating in these provinces. In addition, it is investigated whether a rather rough six-point Likert scale measuring the perception of the institutional setting explains the difference in the performance and thereby eases the analysis for resource-poor countries by lowering the data needs.

Survey Design and Data

A company survey was conducted in three rural provinces of Vietnam (see Figure 12.1 for an overview map). The survey aimed at generating a sample of companies that have a considerable effect on the employment opportunities of the rural population in the provinces and constitute major driving forces of provincial development. The sample frame was therefore set to non-farm companies that employ more than 50 workers and that operate as legal entities in the research province, leading to a sample of 128 companies (36 in Ha Tinh, 46 in TT Hue, and 46 in Dak Lak). Based on the nature of the questionnaire[2] including assessments of the business environment and insight into the basis of business decision making, the interview partners were managers/owners of the companies.[3]

Medium- and especially large-scale companies require well-developed infrastructure and mostly depend on agglomerations (Schätzl, 2003). Therefore, most of the companies operate from the provincial capitals. These companies employ a large share of the non-farm sector workforce also covering rural areas. Table 12.1 gives an overview across the provinces of the shares covered by the survey.

Depending on the source used, the share covered by the survey is at least 17 per cent of the total workforce for Ha Tinh and Dak Lak and 26 per cent for TT Hue and goes as high as 21 per cent in Dak Lak and 42 per cent and 44 per cent for Ha Tinh and TT Hue. As the sample frame or definition of the enterprises used is not provided it is not possible to judge which source would be more accurate for comparison. Nevertheless, the coverage is high enough to be able to generate meaningful results representing the RNFE.

---

2   See Mausch (2010) for the questionnaire used.

3   Fifty-two per cent of the interview partners were the owner of the company, 42 per cent were leading and top managers of the company and the rest were head of several management fields like marketing, trade, or personnel departments.

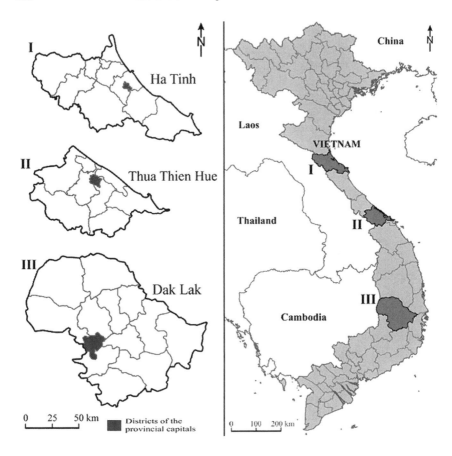

Figure 12.1   Research area – company survey

Source: Own presentation based on Hardeweg et al., 2007.

Table 12.1   Representativeness of the sampled companies

Province	Employees in non-farm enterprises[a]	Employees in enterprises[b]	Employees in sampled companies (N = 128)	Employees covered
	#			%
Ha Tinh	54,413	22,215	8,553	17–42
TT Hue	67,533	40,188	16,712	26–44
Dak Lak	47,428	56,553	8,889	17–21

Source: a: based on GSO, 2008a; b: based on GSO, 2008b.

Table 12.2   Sectoral distribution of companies by province (N = 128)

Sector	Ha Tinh	TT Hue	Dak Lak
Food processing	14%	13%	17%
Garment/textile	0%	15%	2%
Construction	6%	0%	20%
Hotel	6%	11%	2%
Wood processing	17%	4%	11%
Construction material production	8%	20%	0%
Trade	14%	7%	7%
Service	11%	7%	15%
Multiple	14%	7%	15%
Others	10%	16%	11%

Source: Own survey data collected within DFG FOR 756 (2008).

Several differences exist in the sectoral distribution of the companies across the provinces (Table 12.2). First of all, TT Hue has a more 'developed' structure catering to the tourists coming through, a sector that is lacking in the other two provinces, and a textile sector that is virtually absent in the other provinces. The trade and service sector is relatively more important in Ha Tinh with only a few companies involved in production. Additionally, there is also a high share of wood processing. Finally, Dak Lak has a heavy dependence on the construction and food processing sectors. The production and processing of coffee is the major economic factor in Dak Lak province, with almost 40 per cent of the total agricultural land being used for coffee production (GSO, 2008c).

The same diverse picture also holds for the internal structure and characteristics of the companies in the three provinces. Table 12.3 shows that the profits of companies in Ha Tinh are just 15–20 per cent of the profits realized in Dak Lak and TT Hue.

The workforce is not only different in size but also in its structure. In Dak Lak a higher share of both professionals as well as casual labour is employed. A distinct difference is the share of company owners who are members of the communist party of Vietnam. In Dak Lak more than half of the companies are owned by members, whereas in Ha Tinh, which is generally considered a stronghold of the Vietnamese communist party,[4] only 14 per cent of the companies have this link to the ruling party. This fact is likely to be based on the resettlement policy that started after the reunification and continued up to the 1990s. During this time the communist party tried to reverse rural–urban migration. New economic zones were established and people received financial assistance if they decided to migrate to these areas.

---

4   According to personal communication with various experts in Vietnam.

Table 12.3   Surveyed companies' descriptions by province

Variable	Unit	Ha Tinh	TT Hue	Dak Lak	Total
Profit 2006	$0000	113	776	552	506
Employees 2006	#	277	410	237	310
Employment growth 2004–2006	%	45	21	18	27
Workforce composition					
Professionals	%	7	7	11	9
Engineers	%	10	13	11	11
Workers	%	63	67	44	58
Casuals	%	20	13	34	22
Large (>200 employees)	%	36	52	48	46
Party member	%	14	16	59	32
Industrial zone	%	14	27	9	17

Note: All values given are means; N = 128.
Source: Own survey data collected within DFG FOR 756 (2008).

This policy aimed not only to lower the population pressure in the boom regions but also to 'colonize' the central highlands with its high share of ethnic minorities (Wescott, 2003). This system likely favoured party members as the recipients of the assistance available and to reach the goal of colonizing and securing these areas.

The Effect of Institutional Factors on the Performance of the RNFE

Perception of Business Environment and Performance of the Companies

To gain insight into the business environment of the three provinces, companies were asked to give their assessment of the importance and rating of the local conditions considering three major groups of factors, i.e. general business environment, local infrastructure and governmental quality. The assessment was based on a scale ranging from one to six, with six being the best score.[5] Furthermore, companies were asked to indicate up to three of the factors that constitute barriers to expanding their operations. Based on the scores it is possible to assess which constraints these companies face and in which areas political intervention might help to overcome bottlenecks and improve the performance of the economy. Furthermore, it may also point out the factors that need to be considered in order to attract more companies to rural provinces.

---

5   This scale was later normalized to a range from zero to one.

Table 12.4   Perceptions of business environments

	Importance[a]	Rating[a]	Barrier[b]
General business factors			
Skilled labour	0.91	0.58	62
Available credit	0.86	0.72	13
Affordable inputs	0.85	0.35	16
Proximity to supply	0.81	0.47	22
Affordable land	0.74	0.5	8
Proximity to customers	0.72	0.72	17
Affordable labour	0.66	0.56	1
Infrastructure			
Good roads	0.89	0.48	24
Training quality	0.88	0.38	11
Mobile network	0.87	0.76	1
Internet	0.83	0.65	1
Industrial zones	0.79	0.37	10
School quality	0.75	0.62	1
Proximity to airport	0.63	0.6	1
Recruitment agencies	0.58	0.35	1
Governmental quality			
Pro-private business attitude	0.9	0.7	10
Reliability of regulations	0.9	0.57	35
Secure land rights	0.87	0.54	8
Financial gov. support	0.86	0.55	13
Voice	0.83	0.74	0
Non-financial gov. support	0.82	0.52	8
Bribe level	Not applicable	0.64	3

Note: a: all values are means ($\in (0, 1)$ with 1 being the best score); b: number of companies stating factor as major expansion barrier.
Source: Own survey data collected within DFG FOR 756 (2008).

Table 12.4 provides an overview of the companies' responses with factors in each group sorted according to the mean stated importance. Firstly, the general business factors include indicators that are well established as major influences on companies' behaviour and gives skilled labour as the top priority for these companies. This is also related to it being stated that the lack of skilled labour is a major barrier to expansion for almost 50 per cent of the companies. The availability of affordable inputs, which is the third most important factor for these companies, shows the lowest score for the rating of local conditions.

Secondly, the group of the infrastructure endowment of the provinces is given. The most important is the quality of roads which is crucial for any flow of inputs and products and thus for almost all business transactions. Indicating poor road quality, this factor is also rated rather low, and listed as the most important expansion barrier within this group. The two factors, training quality and mobile network, are rated almost equally in terms of importance but the local conditions ratings are considerably different with the mobile network generally rated good and training quality rated very low.

Finally, several governmental quality indicators are considered. The most important factors in this category are the pro-private business attitude of local government officials and the reliability of local regulations. While the pro-private business attitude is rated rather highly, the reliability of the regulations is rated low and is also the second most often stated expansion barrier across all three groups of indicators. One surprising result is that, in contrast to the other two groups, all indicators are considered relatively important. Furthermore, the lowest importance ratings are still better than those found in, e.g., the general business factors. Nevertheless, 45 per cent of the companies interviewed stated at least one of the indicators in this group as one of their three most important expansion barriers. Additionally, the two factors of pro-private business, attitude and reliability of governmental regulations, scored the highest importance values of all factors included.

Disaggregating the expansion barriers by province highlights the fact that companies face very different problems depending on their location. Figure 12.2 gives the number of companies in each province that stated one of the factors within the groups as one of their three major expansion barriers. In each province the group of factors that is most often stated is one of the general business factors. These are the factors that need to be satsified to a sufficient extent for companies to be able to operate reasonably well.

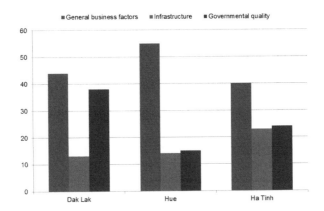

Figure 12.2   Major expansion barriers of companies by province
Source: Own survey data collected within DFG FOR 756 (2008).

The other two groups are more diverse. First, the infrastructure is stated as a problem in almost 50 per cent of the cases in Ha Tinh. Second, the governmental quality was least often mentioned as a problem by TT Hue companies, where it is only slightly more than the infrastructure but both are at a low level. In Ha Tinh it is also very similar to the share of companies stating infrastructure problems but on a much higher level of almost 50 per cent. Finally, in Dak Lak the infrastructure was no problem but governmental quality poses a barrier to almost 60 per cent of the companies.

Based on the details given in Table 12.4, three indices were generated based on the following equation:

$$X_i = \sum_{P=1}^{P} \left[\frac{\sum_{n=1}^{N} R_n * I_n}{N} \right] * \left[\frac{1}{P} \right]$$

Variables:

X$_i$ Index value for the three indices (i)
R$_n$ Rating of each factor (n) included
I$_n$ Importance of each factor (n) included
N Number of factors (n) included in the index
P Number of interviewed companies in province

The indices were developed in order to get a more aggregated picture of the constraints faced by the companies and combine all factor groups into the three group indices, i.e. general business environment, the infrastructure, and governmental quality. Each index combines the stated importance of the factors with the provincial mean rating of the same. As the importance and the ratings are normalized to (0, 1), all parts of the sum also range between zero and one, with one being the best score for both the importance and the rating. In order to again normalize the total indices to a range between zero and one, the sum is divided by the number of factors included. The index weights the rating of each factor by the stated importance of each factor in order to generate an index value that is able to value the impact of the indices on the operations of the companies. This gives more weight to factors that are of great importance to the companies and thus leads to values that reflect the perception of the people who do business in the province. Finally, the mean of the index is taken over the province in which the company is located to be able to evaluate the provincial settings.

Table 12.5 shows the differences in the index scores across the three provinces. The general business environment is rated best in Dak Lak. The score for TT Hue is, although not significantly, higher than for Ha Tinh province. The biggest differences are found in the infrastructure index with the highest value being 0.17 points higher than the lowest value. In this index all provinces perform significantly differently from each other, with TT Hue reaching the highest value and Ha Tinh the lowest.

Table 12.5   Provincial differences in the index scores

		Ha Tinh		TT Hue		Dak Lak	
Business environment	Ha Tinh	–		-0.02		-0.04	*
	TT Hue	0.02		–		-0.06	***
	Dak Lak	0.04	*	0.06	***	–	
Infrastructure	Ha Tinh	–		-0.17	***	-0.15	***
	TT Hue	0.17	***	–		0.02	*
	Dak Lak	0.15	***	-0.02	*	–	
Governmental quality	Ha Tinh	–		-0.10	***	0.01	
	TT Hue	0.10	***	–		0.11	***
	Dak Lak	-0.01		-0.11	***	–	

Note: T-test mean comparison significance levels at the 99% level = ***, 95% level = **, and 90% level = *.
Source: Own survey data collected within DFG FOR 756 (2008).

Finally, the governmental quality index is perceived as significantly better in TT Hue as compared to both of the other provinces, which do not differ significantly from each other. Nevertheless, the mean score for Dak Lak province is the lowest of all provinces.

Effects of the Business Environment Perceptions

Using the indices generated, the effects of these perceptions of the ease of doing business are analysed. These effects are tested on two variables, first the profits of the companies in 2006 and second the size of their workforce in 2006. The regression analysis (see Table 12.6) is implemented in four steps including different sets of variables checking robustness. The first set of variables accounts for some basic business factors and is included in all regressions. These variables are two dummy variables indicating whether the company is large in terms of its workforce[6] and whether the company exports its produce. Furthermore, losses due to natural disasters and the severity of these losses are accounted for by the shock/profit ratio, and the share of replaced permanent workers is included to control for stability and therefore predictability of the business and the reliability of the workforce. Finally, three dummies are included indicating the sector in which the company operates, namely the industry sector, service sector and the food sector. The food sector was separated from the industry sector as it constitutes a large share (30–53 per cent) of the whole industrial group. Furthermore, food

---

6   This factor is only included for the effect on profit as the size of the workforce is the dependent variable in the other regression.

production is of special importance to policy makers especially in developing countries. A dummy variable is included indicating whether the company was recently[7] privatized.

The general factors included in all regressions show two almost universal significant variables. First, exporting companies employ more workers than non-exporting companies as they have a broader customer base and are able to sell to other countries in cases of low demand in some markets. Companies with higher fluctuation rates have a smaller workforce as their operations are less reliable and their focus cannot lie entirely on their operations; they have to deal with the problem of finding new skilled reliable workers. However, this fluctuation does not affect their profits. For employment, the recent private dummy has a positive coefficient throughout, and the food sector dummy has a negative coefficient in all four regressions. The first effect is most likely based on former state-owned companies still having a high share of long-term contracted workers who cannot be laid off even if not required anymore. This is likely to also explain the inconclusive effect on the profits. The second factor suggests that food production involves less labour as compared to the rest of the companies. This sector, including large-scale breweries and coffee processors, operates more mechanization and is therefore less labour-intensive. However, after controlling for the other factors the effect of all three sector dummies becomes insignificant. The other variables included are neither significant in most of the regressions nor do they show a constant sign.

For the profits generated, the dummy for large companies always has a significant positive effect as most of the time bigger companies generate higher profits. For the export and food sector dummies, the signs of the coefficients do not change between the settings and thus the direction of the influence can be assumed positive in the case of exporting and food producing companies.

The second set of variables accounts for several indicators on the institutional setting within which the companies operate. The indicators used are three dummy variables showing whether the company is a member of a business association, a member of the communist party and whether a long-term relationship with the major customer exists. Furthermore, the rating of the local bribe level is included. As outlined before, these factors have a direct impact on the transaction costs the companies face.

Within this set, the number of customers has a significant positive effect on the profits but not on the size of the workforce. Therefore, the reliability and predictability of a broad customer base seems to outweigh the higher transaction costs imposed by dealing with more customers when it comes to profits. However, for the workforce the effect is inconclusive. Insignificant but positive effects emerge from belonging to an association and long-term relations with the major customer.

---

7   Here, recently is defined as less than four years ago, based on the mean number of years the former state owned companies have been private.

Table 12.6 Regression results for profit and employment generation of companies in 2006

Variables	Unit	Results for profit of companies in 2006								Results for number of employees in 2006							
		ey/ex (Rob. S.E.)	Sign	ey/ex (Rob. S.E.)	Sign	ey/ex (Rob. S.E.)	Sign	ey/ex (Rob. S.E.)	Sign	ey/ex S.E.	Sign	ey/ex S.E.	Sign	ey/ex S.E.	Sign	ey/ex S.E.	Sign
Large	dummy	0.438 (0.162)	***	0.539 (0.132)	***	0.472 (0.177)	***	0.534 (0.136)	***								
Exporting	dummy	0.064 (0.174)		0.372 (0.174)	**	0.089 (0.183)		0.340 (0.162)	**	0.523 (0.110)	***	0.418 (0.119)	***	0.511 (0.112)	***	0.411 (0.118)	***
Shock/profit	ratio	-0.123 (0.042)	***	-0.170 (0.051)	***	-0.116 (0.063)	*	-0.130 (0.048)	***	-0.008 (0.032)		-0.003 (0.029)		-0.008 (0.037)		0.014 (0.032)	
Perm. empl. repl.	%	-0.141 (0.131)		0.197 (0.137)		-0.104 (0.144)		0.196 (0.150)		-0.172 (0.057)	***	-0.115 (0.076)		-0.150 (0.061)	**	-0.129 (0.077)	*
Recent private	dummy	-0.162 (0.151)		0.138 (0.172)		-0.276 (0.148)	*	0.082 (0.177)		0.162 (0.096)	*	0.071 (0.102)		0.156 (0.103)		0.052 (0.106)	
Food	dummy	0.120 (0.152)		0.110 (0.117)		0.130 (0.181)		0.133 (0.115)		-0.150 (0.07)	**	-0.091 (0.073)		-0.156 (0.070)	**	-0.090 (0.071)	
Industry	dummy	-0.076 (0.177)		0.192 (0.263)		-0.055 (0.178)		0.133 (0.257)		-0.084 (0.111)		0.005 (0.116)		-0.099 (0.112)		0.010 (0.127)	
Service	dummy	-0.112 (0.109)		-0.010 (0.100)		-0.109 (0.155)		0.012 (0.101)		-0.001 (0.046)		0.034 (0.055)		0.002 (0.047)		0.035 (0.053)	
Association member	dummy			0.201 (0.295)				0.232 (0.289)				0.079 (0.152)				0.125 (0.151)	

Variable	(1)	(2)	(3)	(4)	(5)	(6)	(7)	(8)
Party member (dummy)		-0.145 (0.150)		-0.116 (0.137)		-0.005 (0.083)		-0.023 (0.077)
Long term selling (dummy)		0.232 (0.403)		0.250 (0.390)		0.242 (0.226)		0.253 (0.224)
Customer 80% (#)		0.543 (0.213) **		0.572 (0.214) ***		-0.081 (0.078)		-0.077 (0.076)
Contract defaults (#)		-0.240 (0.136) *		-0.157 (0.154)		-0.193 (0.094) **		-0.157 (0.099)
Bribe level (0,1)		0.128 (0.326)		0.363 (0.363)		-0.313 (0.162) *		-0.265 (0.160) *
Govern-mental quality (0,1)			1.933 (3.982)	3.431 (2.768)			1.117 (1.289)	-0.362 (1.444)
Infra-structure (0,1)			1.741 (1.228)	1.536 (1.210)			0.066 (0.825)	0.948 (0.531) *
Business environment (0,1)	dropped	dropped	dropped	dropped	dropped	dropped	dropped	dropped
N	N = 119	N = 90	N = 111	N = 90	N = 118	N = 90	N = 111	N = 90
F	$F_{(8,110)} = 1.49$	$F_{(14,75)} = 1.86$	$F_{(10,100)} = 1.26$	$F_{(16,73)} = 1.77$	$F_{(7,110)} = 4.86$	$F_{(13,76)} = 2.95$	$F_{(9,101)} = 3.67$	$F_{(15,74)} = 2.68$
Prob > F	Prob>F = 0.1698	Prob>F = 0.0457	Prob>F = 0.2624	Prob>F = 0.0525	Prob>F = 0.0001	Prob>F = 0.0016	Prob>F = 0.0005	Prob>F = 0.0026
R^2	$R^2 = 0.1323$	$R^2 = 0.3277$	$R^2 = 0.179$	$R^2 = 0.3648$	$R^2 = 0.249$	$R^2 = 0.3075$	$R^2 = 0.248$	$R^2 = 0.3182$

Note: Significance at the 99% level = ***, 95% level = **, and 90% level = *; for a detailed description and descriptive statistics of the variables included, see Mausch, 2010.

Source: Own calculations based on own survey conducted within DFG FOR 756 (2008).

The bribe level shows opposite signs in the two sets of regressions. For the profits it shows a positive effect as opposed to the expected negative effect. Thus, in this setting, higher bribe levels lead to greater profits. The explanation for this seemingly contradictory influence might be the anti-corruption policy. On the one hand, the companies that follow the law and do not pay bribes are performing worse and state a low bribe level as in their perspective it is low. On the other hand, the companies that do pay bribes know of the real prevalence of corruption and they benefit from not following the anti-corruption laws and are able to influence the local authorities. For employment generation, the coefficient for the bribe level is significantly negative. Companies that pay less bribes do employ more workers, which might be based on their size as they have a greater influence and are more important for the provincial economy. This might replace bribe payments as their pure size could sufficiently influence the decision makers.

Even though not showing significant coefficients, all other variables in this group have a consistent sign in all regressions. Positive influences emerge from belonging to an association and from long-term relations with the major customer. These effects are aligned with transaction cost arguments, as belonging to an association promotes access to new market partners and increases the influence of members on political decision-making. Long-term relations with the major customer increase the reliability and predictability of the business and enable the company to make long-term plans and therefore employ a larger permanent workforce. Negative effects seem to emerge from party membership, more customers and high incidences of contract defaults. Again, these effects are explained by the transaction cost theory. The more customers a company has, the higher the fluctuation of these customers; relationships are likely to be less regular and therefore less reliable. The same argument of lower predictability also holds for contract defaults. If a company is faced more often with defaults by buyers or by suppliers, the business is more unstable and therefore short-term adjustments need to be made and a lower permanent workforce is employed.

Finally, the three generated indices[8] are included first only with the set of general characteristics and second with all other variables. Out of the three indices included, none shows a significant influence on profits. The governmental quality and the infrastructure evaluation, although not significant, show positive coefficients and are therefore likely to affect profits positively, although the influence cannot be accurately quantified. The infrastructure index shows a significant positive effect in one of the workforce regressions and an insignificant positive effect in the other one. Large companies seem to choose locations with a better infrastructure endowment after controlling for the other relevant factors. Governmental quality shows neither a significant nor a clear direct association

---

8   The index on the general business environment was dropped as the means are very similar across the provinces and using the three indices creates some collinearity problems.

with the company workforce. After controlling for the other relevant factors, the governmental quality shows neither a significant nor a clear directed association with the company workforce.

Summary and Conclusions

Generally, the proxies used to cover the institutional setting do have some explanatory power. Higher numbers of customers have a positive influence on the profits of a company, and smaller companies perceive a higher bribe level than larger companies. The indices generated are only partly able to explain the differences in companies' performances. Nevertheless, this is likely based on the small number of companies/provinces included. This chapter showed that the self-reported perception of institutional factors explains parts of the differences in companies' performances, even though it was only operationalized by using a simple six-point Likert scale for all indicators. Even after accounting for several 'hard' factors on the transaction costs the companies face, the index of their perceptions highlights differences across the provinces beyond these. In contrast to the PCI, face-to-face interviews were utilized, ruling out the problem of a possible self-selection bias in the sample. Furthermore, information on who answered the questions is available, ensuring that all respondents had the same understanding.

The provincial differences emerging from the analysis can partly be attributed to historical factors that cannot be reversed in the short run, e.g. the brain drain that especially Ha Tinh faces. It also illustrates that, like the PCI, there are several important factors concerning governmental quality and the institutional setting that might assist even these unfavoured provinces to generate growth and attract more companies. A generally positive impact was established and the major constraints of availability of skilled labour and the reliability of regulations were shown. In particular, reliability is one factor that can be improved in the short term and would assist in improving the managers' perceptions. The availability of skilled labour seems, at first sight, to be a factor that needs to be addressed by improved schools and universities, but other factors that influence the availability were also obvious. In Ha Tinh especially, reduction of the brain drain is an important aim in making highly skilled workers available to local companies. By creating a fruitful climate for investments and for doing business in general, this problem might be resolved simply by creating attractive jobs locally, instead of centrally in Hanoi and Ho Chi Minh City.

Furthermore, the differences between private companies and recently privatized companies that used to be state-owned emerged not only during the interviews but also came up in the analysis. Companies that were recently privatized appear to face quite different problems as compared to other private companies. Besides having to deal with historical facts such as long-term contracts of workers originating from before privatization, these companies are often run by their former managers and thus changes in operations take more time due to the lack of experience in the private sector. Providing these companies with more assistance in managing the transition would certainly benefit, not only the company itself, but the rural economy in general.

References

ADB, 2005. Vietnam Development Report 2006, in Joint Donor Report to the Vietnam Consultative Group Meeting, 6–7 December, Hanoi.

Coase, R., 1937. The Nature of the Firm. *Economica* 4, 386–405.

Collier, P., 2007. *The Bottom Billion*. New York, Oxford University Press.

Cungo, A., Gow, H., Swinnen, J., et al., 2008. Investment with Weak Contract Enforcement: Evidence from Hungary During Transition. *European Review of Agricultural Economics* 35 (1), 75–91.

Dercon, S., 1996. Risk Crop Choice and Savings: Evidence from Tanzania. *Economic Development and Cultural Change* 44 (3), 485–514.

Dercon, S. and Krishnan, P., 1996. Income Portfolios in Rural Ethiopia and Tanzania: Choices and Constraints. *Journal of Development Studies* 32 (6), 850–75.

DFG FOR 756, 2008. Joint Research Project – Leibniz University of Hannover, Georg-August-University Goettingen, Johann Wolfgang Goethe University Frankfurt am Main, Justus-Liebig University Giessen. For details see: http://www.vulnerability-asia.uni-hannover.de/.

Fafchamps, M., 2004. *Market Institutions in Sub-Saharan Africa: Theory and Evidence*. Comparative Institutional Analysis (CIA) Series. Cambridge, USA, Massachusetts Institute of Technology Press.

Fforde, A., 2001. Light Within the ASEAN Gloom? The Vietnamese Economy Since the First Asian Economic Crisis (1997) and in the Light of the 2001 Downturn. Paper for the Vietnam Update 2001: Governance in Vietnam: The Role of Organizations.

Gagliardi, F., 2008. Institutions and Economic Change: A Critical Survey of the New Institutional Approaches and Empirical Evidence. *Journal of Socio-Economics* 37, 416–43.

Gradstein, M., 2007. Institutional Traps and Economic Growth. Centre for Economic Policy Research – Discussion Paper, 6414.

Grimm, M. and Klasen, S., 2007. Geography vs. Institutions at Village Level. Institute of Social Studies Working Paper, 449.

GSO, 2008a. The Non-Farm Individual Business Establishment Survey. Data provided online by the General Statistics Office of Vietnam: http://www.gso.gov.vn [accessed: 1 January 2008].

——— 2008b. The Real Situation of Enterprises Through the Results of Surveys Conducted from 2002 to 2005. Data provided online by the General Statistics Office of Vietnam: http://www.gso.gov.vn [accessed: 1 January 2008].

————— 2008c. Sections: Statistical Data and Statistical Censuses and Surveys. Various data provided online by the General Stastistics Office of Vietnam: http://www.gso.gov.vn [accessed: 1 January 2008].

Hardeweg, B., Praneetvatakul, S., Tung, P., et al., 2007. Sampling for Vulnerability to Poverty: Cost Effectiveness Versus Precision. Paper to the Conference: Tropentag 2007, Witzenhausen, October 2007.

Krug, B., 1991. Die Transformation der sozialistischen Volkswirtschaften in Zentraleuropa: Ein Beitrag der vergleichenden Oekonomischen Theorie von Institutionen. In: Wagener, H.-J. (ed.), *Anpassung durch Wandel. Evolution und Transformation von Wirtschaftssystemen.* Berlin, Germany, Verein fuer Sozialpolitik, pp. 39–60.

Malesky, E., 2007. The Vietnam Provincial Competitiveness Index 2006 Measuring Economic Governance for Private Sector Development. Hanoi, Vietnam, VNCI.

Malesky, E. and Ray, D., 2007. The Vietnam Provincial Competitiveness Index 200 Measuring Economic Governance for Private Sector Development Summary Report. Hanoi, Vietnam, VNCI.

Mausch, K., 2010. *Poverty, Inequality and the Non-Farm Economy: The Case of Rural Vietnam.* Berlin, Logos publishing house.

Rao, P., 2003. *The Economics of Transaction Costs – Theory, Methods and Applications.* Basingstoke, United Kingdom, Palgrave Macmillan.

Reardon, T., 1997. Using Evidence of Household Income Diversification to Inform Study of the Rural Nonfarm Labor Market in Africa. *World Development* 25 (5), 735–47.

Revilla Diez, J., 1995. *Systemtransformation in Vietnam: Industrieller Strukturwandel und regionalwirtschaftliche Auswirkungen.* Hannover, Hannoversche Geographische Arbeiten, p. 51.

Sala-i-Martin, X., 2006. The World Distribution of Income: Falling Poverty and … Convergence, Period. *The Quarterly Journal of Economics* 121, 351–97.

Schätzl, L., 2003. *Wirtschaftsgeographie 1 – Theorie.* Vol. 9. Paderborn, Germany, Schoeningh publishing house.

Schupp, F., 2002. Growth and Inequality in South Africa. *Journal of Economic Dynamics and Control* 26, 1699–720.

Shankar, R. and Shah, A., 2001. Bridging the Economic Divide Within Nations: A Scorecard on the Performance of Regional Development Policies in Reducing Regional Income Disparities. World Bank Policy Research Working Paper, 2717.

Transparency International, 2008. Corruption Perception Index 2008. Available at: http://www.transparency.org./content/download/36450/573413 [accessed: 12 February 2015].

United Nations, 2000. United Nations Millenium Declaration. General Assembly, 18 September, 55/2.

————— 2005. The Inequality Predicament Report on the World Social Situation 2005. New York, United Nations.

———— (ed.), 2007. Persistent and Emerging Issues in Rural Poverty Reduction. Bangkok, Thailand, Economic and Social Commission for Asia and the Pacific.

Venables, A., 2003. Spatial Disparities in Developing Countries: Cities, Regions and International Trade. *Journal of Economic Geography* 5 (1), 3–21.

Wescott, C., 2003. Hierachies, Networks, and Local Government in Vietnam. *International Public Management Review* 4 (2), 20–40.

World Bank, 2003. Vietnam Development Report 2004. Poverty Reduction and Economic Management Unit – East Asia and Pacific region, Washington DC, World Bank.

———— 2007. Will Resilience Overcome Risk? Special Focus: Agriculture for Development. East Asia and Pacific update, Washington DC, World Bank.

PART IV
Conclusion

Conclusion: Perspectives and Outlook on Rural Development

Peter Dannenberg and Elmar Kulke

The case studies and conceptual approaches presented in this book demonstrate a large variety of complex challenges and structural deficits (see also Anríquez and Stamoulis, 2007; Sedlacek et al., 2009) that exist for rural areas in an era of fluctuating local and global dynamics. However, the studies also reveal a fresh variety of options and strategies for regional actors and stakeholders to cope with these challenges and, in some cases, to develop proactive approaches for economic and regional development and to overcome or prevent a negative cumulative process of increasing structural deficits (as outlined e.g. by Wießner, 1999).

Dynamics in Food Production

The results presented in Part II suggest that there are options available even to small scale farmers in developing countries, who are often among the most marginalized and vulnerable members of society. Dannenberg and Nduru (Chapter 2) showed that the establishment of and the integration in interlinked regional production systems (see also Dannenberg and Kulke, 2005) increases the chances of all stakeholders – including those of small-scale farmers – to stay competitive even in sophisticated international value chains which are driven by large retail companies (e.g. supermarkets). These findings are in line with the conclusions drawn by Vishwanath Gowdru et al. (Chapter 3), who found that group standard certification and collective marketing (as forms of upgrading), together with political support, additional technical training, and market information systems increases the chances of small farmers to supply organic and conventional food markets in India.

Regarding the global North our studies underline the importance for regional agrarian systems to stay flexible and to adjust the ways of doing business to ongoing dynamics. The study conducted by Miller et al. (Chapter 4) on farming systems in the United States and Belgium outlines potential weaknesses of regional agrarian systems if they only focus on themselves. However, the study also demonstrates methods used by farmers to overcome the lock-in effects of traditional farming technologies and practices. Jablonski (Chapter 5) expands on this further, discussing the possibilities that re-localized food systems present a viable strategy to support rural economic development. Voigt and Mergenthaler (Chapter 6) contribute to the discussion by outlining typical success factors, bottlenecks and starting points for policy advice.

The case studies addressed in Part II provided fertile ground for the critical discussion that followed about the possible strategies that farmers and regional food production systems can adopt to remain integrated in food value chains and/ or enter new markets. Furthermore, they also identified different ways in how far the national and local economic, social and institutional frameworks can support such strategies.

Scientists and universities who work directly with the actors and stakeholders involved or provide a general knowledge base for policies, planning, and businesses can also play an important role in these processes.

Alternative Functions of Rural Areas

This is also true for the identification and critical examination of alternative functions of rural areas beyond food production (Part III). Here Panzer-Krause (Chapter 7) outlined the challenges of rural renewable energy producers faced in accessing financial capital in Germany. The study presented by Trebbin, Franz, and Hassler (Chapter 8) illustrates how cooperation between wholesalers can contribute to secure local supply for local retail infrastructure in rural areas.

Kacprzak and Maćkiewicz (Chapter 9) and Kołodziejczak (Chapter 10) explored how various environmental, agricultural and touristic functions of rural areas and farming can fit together for sustainable rural development, identifying different obstacles and success factors. Heffner and Czarnecki (Chapter 11) provided evidence that the growth of special forms of housing – in their case second homes in Poland – can significantly contribute to the economic and social development of rural areas.

The example of small and medium enterprises (SMEs) in rural Vietnam presented by Mausch and Diez (Chapter 12) indicates the general potential of off-farm employment even in rural areas of the global South (see also Brünjes and Diez, 2012). However the study also identified significant challenges and constraints for such companies which included in the shortage of skilled workers, adequate infrastructure to connect to the markets, and reliable and pro-private business orientated regulation by the regional governments.

Summing up, the case studies outlined do not only contribute to further elaborate and refine the state of the art on agricultural geography and rural development. They also showed how and to what extent such scientific work can help to derive applied and political recommendations which help to support the actors in rural areas in dealing with the local and global challenges outlined.

References

Anríquez, G. and Stamoulis, K., 2007. Rural Development and Poverty Reduction: Is Agriculture Still the Key? *Journal of Agricultural and Development Economics* 4, 5–46.

Brünjes, J. and Diez, J.R., 2012. Opportunity Entrepreneurs: Potential Drivers of Non-Farm Growth In Rural Vietnam? Working Papers on Innovation and Space 1.

Dannenberg, P. and Kulke, E., 2005. The Importance of Agrarian Clusters for Rural Areas: Results of Case Studies in Eastern Germany and Western Poland. *Die Erde* 136, 291–309.

Sedlacek, S., Kurka, B., and Maier, G., 2009. Regional Identity: A Key to Overcome Structural Weaknesses in Peripheral Rural Regions? *European Countryside* 1, 180–201.

Wießner, R., 1999. Ländliche Räume in Deutschland – Strukturen und Probleme im Wandel. *Geographische Rundschau* 51, 300–304.

Index

Printed and bound by CPI Group (UK) Ltd, Croydon, CR0 4YY

22/10/2024

01777628-0015